Springer Series on
Atoms+Plasmas

Editor: J. P. Toennies

11

Springer Series on
Atoms+Plasmas

Editors: G. Ecker P. Lambropoulos I. I. Sobelman H. Walther
Managing Editor: H. K. V. Lotsch

V. I. Lengyel   V. T. Navrotsky
E. P. Sabad

Resonance Phenomena
in Electron-Atom Collisions

With 64 Figures

Springer-Verlag
Berlin Heidelberg New York
London Paris Tokyo
Hong Kong Barcelona
Budapest

Professor Dr. Vladimir I. Lengyel
Dr. Vyacheslav T. Navrotsky
Uzhgorod University, Oktyabrskaya Street 32, 294000 Uzhgorod, Ukraine

Dr. Emil P. Sabad
Institute for Nuclear Research of the Ukrainian Academy of Sciences, Uzhgorod Branch,
Mukatchevskaya Street 69, 294000 Uzhgorod, Ukraine

Series Editors:

Professor Dr. Günter Ecker
Ruhr-Universität Bochum, Fakultät für Physik und Astronomie,
Lehrstuhl Theoretische Physik I, Universitätsstrasse 150,
W-4630 Bochum 1, Fed. Rep. of Germany

Professor Peter Lambropoulos, Ph. D.
University of Crete, P.O. Box 470, Iraklion, Crete, Greece, and
Department of Physics, University of Southern California, University Park,
Los Angeles, CA 90089-0484, USA

Professor Igor I. Sobelman
Lebedev Physical Institute, USSR Academy of Sciences,
Leninsky Prospekt 53, 117333 Moscow, Russia

Professor Dr. Herbert Walther
Sektion Physik der Universität München, Am Coulombwall 1,
W-8046 Garching/München, Fed. Rep. of Germany

Guest Editor: Professor Dr. J. Peter Toennies
Max-Planck-Institut für Strömungsforschung, Bunsenstrasse 10,
W-3400 Göttingen, Fed. Rep. of Germany

Managing Editor: Dr. Helmut K.V. Lotsch
Springer-Verlag, Tiergartenstrasse 17, W-6900 Heidelberg, Fed. Rep. of Germany

ISBN-13:978-3-642-84518-5   e-ISBN-13:978-3-642-84516-1
DOI: 10.1007/978-3-642-84516-1

Library of Congress Cataloging-in-Publication Data. Len'del, V. I. [Teoriia rezonansnykh iavleniĭ v ėlektron-atomnykh stolknoveniĭakh. English] Resonance phenomena in electron-atom collisions / V. I. Lengyel, V. T. Navrotsky, E. P. Sabad. p. cm. – (Springer series on atoms & plasmas ; v. 11) Translation of: Teoriia rezonansnykh iavleniĭ v ėlektron-atomnykh stolknoveniĭakh. Includes bibliographical references and index. ISBN-13:978-3-642-84518-5 1. Electron-atom collisions. 2. Nuclear magnetic resonance. I. Navrotsky, V. T. (Vyacheslav T.), 1954–. II. Sabad, E. P. (Emel 'iăn Petrovich) III. Title. IV. Series. QC793.5.E628L4613 1992 539.7'57–dc20 91-42475

© Springer-Verlag Berlin Heidelberg 1992 ·
Softcover reprint of the hardcover 1st edition 1992

Typesetting: Springer T_EX in-house system
54/3140 - 5 4 3 2 1 0 – Printed on acid-free paper

Preface

Resonance phenomena have been the topic of a number of reviews, and separate questions have been elucidated in some monographs. But the absence of a balanced integral account of the current status of the problem hinders the orientation in this area. The present book is an attempt to fill this gap.

The results of investigations of the resonance scattering of electrons by atoms and ions are considered. We compare different theoretical methods of description of resonance phenomena, for example, the close-coupling method, R-matrix method, and diagonalization method. Special attention is paid to the analysis of the accuracy of the theoretical calculations and experimental data. Besides the conventional analytical solutions of a multiparticle problem, more recently developed methods, made possible by high speed computers, are discussed in detail. Several computer programs are scrutinized.

This book is intended for physicists engaged in the problems of electronic and atomic collisions, and related areas such as plasma and laser physics. It should be of interest to university students and postgraduates.

The authors are grateful to M.I. Gaysak and S.Yu. Medvedev for lengthy discussions, V.V. Balashov, V.B. Belyayev, M.K. Gailitis, R.Ya. Damburg, G.F. Drukarev, B.N. Zakhar'ev, I.I. Fabrikant and G.F. Filippov for their comments on the original results presented here, and O.I. Zatsarinny for fruitful conversations concerning computational problems. We are also grateful to Dr. A.V. Snegursky for his help in translating the text. We are particularly indebted to Professor J.P. Toennies for careful reading of the manuscript and for many helpful suggestions.

Uzhgorod, February 1992

V.I. Lengyel
V.T. Navrotsky
E.P. Sabad

Contents

1. Introduction

Scattering of electrons by atoms and ions has been attracting considerable interest during the last few decades. The reason is that an understanding of elementary processes occurring in the collisions of electrons with atoms and ions is critical in the further development of plasma physics, laser technology, fusion energetics, quantum chemistry, astrophysics, physics of the upper atmosphere, physics of nuclear reactions involving heavy ions. For example, atom and ion excitation by electron impact is one of the most widely used methods of gas laser pumping; most of the emission lines observed by astrophysicists are generated by the electron impact excitation of positive ions, among other particles. In these processes, an important role is played by quasistationary Autoionizing States (AIS) consisting of an incident electron and a target the Auger decay of which leads to the complex resonance structure in the scattering cross sections. These states are atomic analogs of the compound states of nuclei well known in nuclear physics.

In this introductory chapter, we shall give a brief account of the history of the topic and define the scope of the book.

1.1 General Overview

Besides the importance of AIS in electron–atom and electron–ion scattering, they play a considerable role in plasmas and solids. The collisions of ions and atoms in a plasma lead to the intense excitation of multiply charged ionic AIS which, accordingly, affects the energy balance and serves as a convenient diagnostic tool. Another plasma process in which AIS are important is recombination. In solids, when the conductive band electrons interact with impurity atoms in a crystalline lattice an AIS may form. Resonances have also been observed in nuclear processes, in the scattering of elementary particles, and in all these cases the appearance of resonances signifies that the target has a complex structure, e.g., for elementary particles, their quark structure.

At the 3rd International Conference on Electron and Atom Collisions in 1963, H. S. W. Massey ranked the discovery of these resonances among the most exciting achievements in atomic physics of the previous two decades. Since then theoretical and experimental studies of resonances have remained a significant trend in modern atomic physics [1.1–19].

The first indications of the existence of AIS were obtained by J. Frank and co-workers as far back as 1921 in their studies of electric discharges in neon. To explain the data, Frank suggested that when an outer-shell electron of a neon atom is excited into an unfilled orbit, capture of the incident electron is possible, which leads to the formation of an excited negative neon ion.

Later, similar phenomena were also discovered in other atoms. Thus *Whiddington* and *Priestley* [1.20], in a study of electron impact ionization of the He atom, discovered a sharp peak in the scattered electron energy loss spectra in the vicinity of 60 eV. It was, in fact, the first direct indication of the existence of a resonance in the total cross section of electron-helium scattering. This resonance was attributed to the occurence of the He AIS with the 2s2p $^1P^0$ configuration. At the same time Pashen and Kruger, in investigating the Be and Mg vacuum ultraviolet spectra (VUV), discovered atomic states lying above the ionization thresholds. However, because of the difficulties in obtaining high density metal vapors and because of the lack of powerful sources of continuous radiation in the VUV region and intense sources of monochromatic electrons, as well as because of the lack of any application-driven interest, these studies were not pursued further until much later.

Systematic experimental studies of resonances were resumed only in the early 1960s since by that time the energy resolution of electrostatic analysers had achieved the level of several dozen meV at quite acceptable currents (about 10^{-7}A) and the utilization of the synchrotron radiation of electron accelerators as high intensity continuous for VUV radiation sources was successfully realized.

The *Fano*'s pioneering theoretical paper [1.21] facilitated the revival of interest in resonances in atomic systems. Fano showed that the intereference of a discrete state with the adjacent continuum produces resonances in the elastic and excitation cross sections. However, obtaining numerical values required extensive computations which became possible only on high-speed computers. Methods were developed which allowed the calculation of resonances in electron-hydrogen atom scattering even before the appearance of the experimental data. For example, *Burke* and *Shey* [1.22] had discovered the existence of a resonance in elastic electron scattering by a hydrogen atom below the excitation threshold. The explanation for the production mechanism of this resonance was suggested by *Gailitis* and *Damburg* [1.23].

The first resonance discovered experimentally was the ^2S resonance observed by *Schulz* [1.24] in elastic e–He scattering just about 0.5 eV below the 2^3S threshold.

Later, due to a significantly increasing interest, resonances were found in electron scattering by atoms and ions of inert gases [1.25, 26], alkali [1.27, 28] and alkaline-earth elements [1.29, 30], mercury atoms [1.31] and some molecules [1.32]. Since then an immense number of papers devoted to this problem have appeared. One may be easily convinced of the significance of this field by studying the Proceedings of the 13th (Berlin, 1983), 14th (Stanford, 1985), 15th (Brighton, 1987) and 16th (New York, 1989) International Conferences on the Physics of Electron and Atom Collisions.

1.2 Scope of the Book

Chapter 2 deals with the main concepts concerning potential scattering which are essential for understanding the following material. The principal methods of description of electron scattering by a target of known structure are considered in Chap. 3 on the example of a hydrogen atom. The methods of determination of the wave functions and the energies of the AIS of two-electron systems as well as the two-channel theory of resonance scattering are discussed in Chap. 4. Chapter 5 is devoted to the description of a simplified method of calculation of resonances – the Diagonalization Method (DM) – using an example of elastic electron scattering by a helium ion. A multichannel theory is discussed in Chap. 6. Here we have put the emphasis on the application of DM to multichannel scattering. Finally, Chap. 7 surveys the more recent experimental and theoretical results on the scattering of electrons by atoms and ions.

2. Scattering by a Potential Field

The present chapter gives a brief presentation of the material on the problem of scattering by a potential field. This information is essential both to understand the subsequent interpretation and to avoid the necessity of referring to the basic textbooks. Scattering by a potential is a basis for real electron scattering by atoms. It turns out that even in this approximation certain resonances may occur, whose brief description will be presented in this chapter. The experienced reader can skip this material.

2.1 Formulation of the Problem

Particle scattering is one of the best sources of data on the microscopic properties of matter. In the nonrelativistic case, the Schrödinger equation is the starting point of scattering theory.

Consider the most simple case of the scattering of a pair of spinless and structureless nonrelativistic particles. The Schrödinger equation for these two particles is

$$i\hbar\frac{\partial \Psi}{\partial t} = -\frac{\hbar^2}{2m_1}\Delta_1\Psi - \frac{\hbar^2}{2m_2}\Delta_2\Psi + V(r_1, r_2)\Psi , \qquad (2.1)$$

where $\Psi \equiv \Psi(r_1, r_2)$ is the wave function of the system of two particles, $V(r_1, r_2)$ is their interaction potential, Δ_1 and Δ_2 are kinetic energy operators of the first and second particles, respectively.

The wave function Ψ contains all of the information about the scattering processes. Hence the problem is reduced to the solution of (2.1) at a given $V(r_1, r_2)$ and a positive total energy of the system. The solution of (2.1) must satisfy the following boundary conditions: at $r \rightarrow 0$ the solution is finite, while at $r \rightarrow \infty$ it is the sum of the scattered and incident waves. In the case under consideration the interaction energy depends only on the relative position of the particles, i.e.,

$$V(r_1, r_2) = V\left(|r_1 - r_2|\right) . \qquad (2.2)$$

This condition allows a considerable simplification of the initial Eq. (2.1).

Let us introduce the new coordinates

$$r = r_1 - r_2, \quad R(m_1 + m_2) = m_1 r_1 + m_2 r_2 \tag{2.3}$$

and denote

$$m_1 + m_2 = M . \tag{2.4}$$

We present Ψ in the form

$$\Psi(r_1, r_2, t) = \Psi(R)\Psi(r)\exp(-iE_1 t/\hbar) , \tag{2.5}$$

where E_1 is the total energy of the system. The substitution of (2.5) into (2.1) produces

$$-\frac{\hbar^2}{2m}\frac{\Delta_r \Psi(r)}{\Psi(r)} - E + V(r) = 0 ;$$
$$-\frac{\hbar^2}{2M}\frac{\Delta_R \Psi(R)}{\Psi(R)} - E_2 = 0 , \tag{2.6}$$

where E and E_2 are some constants.

It is seen that the second equation in (2.6) describes free motion. This equation contains no information on the interaction and is of no interest to us. Thus the problem is reduced to the solution of the Schrödinger equation

$$-\frac{\hbar^2}{2m}\Delta \Psi(r) - [E - V(r)]\Psi(r) = 0 , \tag{2.7}$$

where $E = E_1 - E_2$ is the energy of the relative motion of the system. Equation (2.7) describes the scattering of a single particle with reduced mass $m = m_1 m_2/(m_1 + m_2)$ in the external potential field $V(r)$. Scattering of this type is, for brevity, called potential scattering.

If one introduces the wave vector $k, (p = \hbar k$, where p is the momentum of the particle), then $E = \hbar^2 k^2/2m$. In terms of k^2, the Schrödinger equation has the form

$$\left[\Delta + k^2 - \frac{2m}{\hbar^2}V(r)\right]\Psi(r) = 0 . \tag{2.8}$$

The solution of (2.8) depends on the boundary conditions. In this chapter we shall consider the potential V which decreases with $r \to \infty$ more rapidly than $1/r^2$.

The form of the asymptotics of the solution of (2.8) depends on the particular formulation of the problem. In scattering experiments, the collimators (Fig. 2.1) select the planar monochromatic wave falling along the Z-axis, θ is the scattering angle. Since at infinity, the scattered spherically diverging wave must be present too, the function Ψ must have the asymptotic form [2.1]

$$\Psi(r \to \infty) = e^{ikz} + f(k, \theta)\frac{e^{ikr}}{r} . \tag{2.9}$$

5

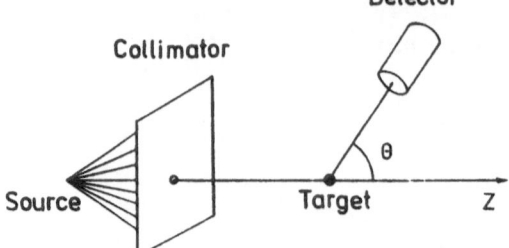

Fig. 2.1. Layout of a scattering experiment

Detector

Collimator

Source

Target

θ

Z

Thus, the problem is reduced to the search for the solution of (2.8) (finite at zero) for the continuous spectrum of $k^2 \geq 0$ values satisfying at $r \to \infty$ the boundary condition (2.9). Hereafter, we shall use the system of atomic units ($\hbar = m = e = 1$).

All of the information on the scattering is contained just in the $f(k, \theta)$ function which is called the scattering amplitude and is exactly the object of the present study.

2.2 Solution of the Schrödinger Equation for Potential Scattering

Let us transfer the last term of the left-hand side of (2.8) to the right-hand side and consider it as a known function. Then the solution of the "inhomogeneous" equation obtained will be sought as its expansion over the solutions of the homogeneous equation, i.e., in the following form:

$$\Psi(r) = \int dk' a(k') \exp(ik'r) . \tag{2.10}$$

The $a(k)$ may be found by the substitution of (2.10) into the left-hand side of the inhomogeneous equation. The solution of the inhomogeneous equation may be obtained by the substitution of $a(k)$ into (2.10) and by adding the solution of the homogeneous one

$$\Psi(r) = A \exp(ikz) + 2 \int V(r')\Psi(r')G(r, r')dr' , \tag{2.11}$$

where

$$G(r, r') = \frac{1}{(2\pi)^3} \int \frac{\exp\left[ik'(r - r')\right] dk'}{k^2 - k'^2} . \tag{2.12}$$

Expression (2.11) is not, in fact, a solution, but turns out to be an integral equation for Ψ which is equivalent to the differential (2.8) and is convenient because of the boundary conditions (2.9) are taken there into account explicitly.

In this case the Green's function (2.12) can be represented in the form [2.1]

$$G(\mathbf{r}, \mathbf{r}') = -\frac{1}{4\pi} \frac{\exp(ik|\mathbf{r} - \mathbf{r}'|)}{|\mathbf{r} - \mathbf{r}'|} .$$

(2.13)

Then the solution (2.11) may be written as

$$\Psi(\mathbf{r}) = e^{ik_i r} - \frac{1}{4\pi} \int \frac{\exp(ik|\mathbf{r} - \mathbf{r}'|)}{|\mathbf{r} - \mathbf{r}'|} U(\mathbf{r}')\Psi(\mathbf{r}')d\mathbf{r}' ,$$

(2.14)

where k_i is the vector directed along the z-axis.

Since the dimension of the scatterer is far smaller than the distance r to the screen or detector, one may write

$$\frac{\exp(ik|\mathbf{r} - \mathbf{r}'|)}{|\mathbf{r} - \mathbf{r}'|} \approx e^{ikr - ik_r r'} ,$$

(2.15)

where k_r is the vector whose value is equal to $|k|$ in the direction r. Then comparison of (2.14) with (2.9) shows that the scattering amplitude is

$$f(k, \theta) = -\frac{1}{4\pi} \int e^{-ik_r r'} U(\mathbf{r}')\Psi(\mathbf{r}')d\mathbf{r}' ,$$

(2.16)

where

$$U(r) = 2V(r) .$$

(2.17)

It is seen that the scattering amplitude is expressed via the solution $\Psi(r)$ of the Schrödinger equation. Further we will use the term "interaction potential" to mean $U(r)$.

Recently, direct methods for determining the scattering amplitude based on the numerical solution of the Schrödinger equation have been developed. These methods will be the central issue of this book.

The most important experimental result of a scattering process is the quantity

$$\frac{d\sigma}{d\Omega} = |f(k, \theta)|^2 ,$$

(2.18)

the differential scattering cross section. An integration of $d\sigma/d\Omega$ over the solid angle Ω gives the total integral scattering cross section

$$\sigma = 2\pi \int_{-1}^{+1} d(\cos \theta)|f(k, \theta)|^2 .$$

(2.19)

Experimentalists determine the differential cross section as the quotient $1/N_a[(n/\Omega)/(N/S')]$, where n and N are the numbers of scattered and incident particles, respectively. Since the solid angle $\Omega = S/r^2$ (here S is the detector area, r is the distance from the target to the detector), and the number of target atoms is equal to $N_a = S'\varrho l$, where l and S' are the thickness and the area of target, respectively; ϱ is the number of scattering centers per unit volume, then $d\sigma/d\Omega$ can be expressed

$$\frac{d\sigma}{d\Omega} = \frac{n}{N}\frac{r^2}{S\varrho l} \tag{2.20}$$

where the amplitude does not depend on the azimuthal angle ϕ. Thus, it follows from (2.19) that the total scattering cross section is determined as the ratio of the flux scattered over all angles to the incident flux. The measurement of differential cross sections at different angles is a very difficult task. That is why experimentalists have usually measured the total cross section by means of a beam attenuation technique. In this method the incident flux j_{in} and the flux of the particles that have *not* been scattered j is detected. The beam attenuation j/j_{in} is related to the total cross section by

$$j = j_{in}\exp(-\varrho l\sigma) , \tag{2.21}$$

Note that these considerations are correct only when forward scattering is negligible and the target thickness is low enough so that only single collisions of the incident particles occur.

2.3 The Radial Schrödinger Equation

Above we have expressed the experimental observables through the solutions of the total Schrödinger equation (2.8) in a 3-dimensional coordinate system. In the low-energy region, reduction to the system of ordinary differential equations by expanding the wave function Ψ in a series of partial waves is the usual method of the solution.

It is well known that if the potential V has spherical symmetry, i.e., depends only on r, the solution can be presented in the form

$$\Psi(\boldsymbol{r}) = \sum_{l=0}^{\infty} C_l R_l(r) P_l(\cos\theta) . \tag{2.22}$$

The substitution of (2.22) into the Schrödinger equation (2.8) and the introduction of the new function

$$U_l(R) = r R_l(r) \tag{2.23}$$

produces a new equation

$$U_l''(r) + \left[k^2 - U(r) - l(l+1)/r^2\right] U_l(r) = 0 . \tag{2.24}$$

Since it is obvious from the physical considerations that $R_l(r)$ must remain finite as $r \to 0$, it follows from the very definition of $U_l(r)$ that $U_l(0) = 0$. Further, we let $U(r)$ decrease more rapidly than the centrifugal potential $l(l+1)/r^2$. Then as $r \to \infty$, (2.24) is reduced to an equation in spherical Bessel functions whose solution may be represented as

$$U_l(r) \approx kr \left[a_l j_l(kr) + b_l n_l(kr)\right] . \qquad (2.25)$$

By using the asymptotics of Riccati-Bessel functions, (2.25) can be rewritten

$$U_l(r) \sim \sin(kr - \pi l/2 + \delta_l) , \quad U_l(0) = 0 , \qquad (2.26)$$

and the phase shift δ_l is related to the scattering amplitude

$$f(k, \theta) = \sum_{l=0}^{\infty} (2l + 1) f_l P_l(\cos \theta) , \qquad (2.27)$$

where

$$f_l(k) = \left(e^{2i\delta_l} - 1\right) / 2ik \qquad (2.28)$$

is the partial amplitude.

The search for $f_l(k)$ is reduced to a determination of the phase (of phase shift) δ_l. Knowledge of δ_l allows one to recover the scattering amplitude according to (2.28), which, in turn, allows one to obtain the experimentally observable cross section of the scattering.

2.4 Numerical Methods for the Solution of the Schrödinger Equation

Direct numerical methods of solving the Schrödinger equation for potential scattering problems have become the most widely used since the advent of ever faster computing systems. This is particularly true for scattering of complex particles.

Let us consider some properties of the solution of the radial Schrödinger equation for potential scattering. Numerical solution of (2.24) has difficulties: i) setting of the sought-for function on the ends of intervals is inconvenient. To formulate the numerical procedure, one must set the value of $U_l(r = 0)$ and the derivative $U_l'(r = 0)$ which is unknown; ii) otherwise the asymptotic value of $U_l(r \to \infty)$ is unknown since it contains the phase δ_l which must be determined. Thus, one has to seek the partial solution of (2.24) at arbitrary boundary values. They may be written as

$$U_l^{(1)}(r = 0) = 0 ; \quad U_l'^{(1)}(r = 0) = 1 . \qquad (2.29)$$

A general solution is determined by means of this partial one. That is, one defines two linearly independent partial solutions whose linear combination yields the general solution

$$U_l(r) = C_1 U_l^{(1)}(r) + C_2 U_l^{(2)}(r) . \qquad (2.30)$$

9

Since the behaviour of the wave function as $r \to 0$ is determined by the centrifugal potential $l(l+1)/r^2$, at zero these two solutions behave as r^{l+1} and r^{-l}, respectively. Since the unknown solution should be regular at zero, $C_2 = 0$. Thus one only has to determine C_1. Two different ways exist which depend on the $U_l (r \to \infty)$ asymptotics. Let us seek the partial solution U_l with boundary conditions (2.26). We numerically integrate (2.24) until the influence of the interaction potential becomes negligible. Let us suggest that this occurs at some $r = R$ (say, near 10 Bohr radii). Then one can assume that the asymptotic region is already reached and only the matching of the numerical solution to the asymptotic one must be carried out. We assume that at the matching point not only the solutions but also their derivatives coincide. In other words, the following equations are assumed to be valid

$$C_1 U_l^{(1)}(R) = \sin(kR - \pi l/2 + \delta_l) \; ;$$
$$C_1 U_l'^{(1)}(R) = k \cos(kR - \pi l/2 + \delta_l) \; . \tag{2.31}$$

The constant C_1 can be excluded from this system of equations and the phase δ_l can be determined from

$$k \cot(kR - \pi l/2 + \delta_l) = U_l'^{(1)}(R)/U_l^{(1)}(R) \; . \tag{2.32}$$

Other forms of U_l asymptotics are possible:

$$U_l(r \to \infty) = \sin(kr - \pi l/2) + \tan \delta_l \cos(kr - \pi l/2) \; , \tag{2.33a}$$

$$U_l(r \to \infty) = i^l \frac{\sin(kr - \pi l/2)}{k} + f_l \exp(ikr) \; , \tag{2.33b}$$

$$U_l(r \to \infty) = \exp(-ikr) + S_l \exp(ikr) \; ; \tag{2.33c}$$

all of them are equivalent. The quantity $K_l = \tan \delta_l$ is called the K-matrix. In (2.33b) $f_l = k^{-1} \exp(i\delta_l)$ is the partial scattering amplitude and $S_l = \exp(2i\delta_l)$ in (2.33c) is called the S-matrix.

It should be noted that in the case of potential scattering the K_l, f_l and S_l quantities are not, in fact, matrices. They will become matrices only if the scattering occurs from objects having some structure (Chap. 3). The K-matrix is the most suitable in our case because we deal with clearly defined real values, which is most convenient for numerical calculations. Thus, instead of the system (2.31) we have

$$C_1 U_l^{(1)}(R) = \sin(kR - \pi l/2) + \tan \delta_l \cos(kR - \pi l/2) \; ,$$
$$C_1 U_l'^{(1)}(R) = k \cos(kR - \pi l/2) - \tan \delta_l k \sin(kR - \pi l/2) \; . \tag{2.34}$$

Solving (2.34) we obtain

$$\tan \delta_l = \frac{U_l^{(1)}(R)k \cos(kR - \pi l/2) - U_l'^{(1)}(R) \sin(kR - \pi l/2)}{U_l^{(1)}(R)k \sin(kR - \pi l/2) + U_l'^{(1)}(R) \cos(kR - \pi l/2)} \; . \tag{2.35}$$

Thus, the problem is reduced to finding the solution with simple boundary conditions (2.29). To derive the solution numerically we use a Taylor series expansion. The integration region is divided into n intervals with mesh width Δr. On the first step we have $U(\Delta r_1) = U(0)+U'(0)\Delta r$, on the second $U(\Delta r_2) = U(\Delta r_1)+U'(\Delta r_1)\Delta r$, etc. Here $U'(\Delta r) = U'(0)+U''(0)\Delta r$ and the initial values of $U(0)$ and $U'(0)$ are taken from the boundary conditions; $U''(0)$ may be obtained from the Schrödinger equation.

The integration is performed up to the value of R at which the influence of the potential becomes negligible. The phase shift is determined from (2.35). In practice, the more precise Numerov, Milne, Runge-Kutta and other methods are used.

It is necessary, however, to consider one more complication related to the singular terms in (2.24). The potentials U which we will meet in the future are generally of the form $U(r) \sim (2Z/r)\exp(-ar)$ (a screened Coulomb potential), i.e., they are singular as $r \rightarrow 0$. Moreover, the case with $l \neq 0$ produces a new difficulty in the numerical computation due to the term $l(l+1)/r^2$ which diverges as $r \rightarrow 0$. In this case solutions of the Bessel-function type are used frequently. For the sake of convenience we shall look for the solution at low r in the form

$$U_l(r) = r^\sigma(a_0 + a_1 r + a_2 r^2 + \dots) . \tag{2.36}$$

The substitution of this solution into the initial equation, where the first term of the exponential expansion is retained gives $\sigma = l + 1$ and $\sigma = -l$. The second value corresponds to the nonregular solution which must be omitted. For the expansion coefficients we obtain the recurrence relation

$$a_{k+1} = \frac{-2Za_k}{(l+k+1)(l+k+2) - l(l+1)} . \tag{2.37}$$

The boundary conditions (2.26) hold provided $a_0 = 1$. Then the numerical solution may be started not from $r = 0$ but, say, from $r = 0.01$. From (2.36) we obtain that $U_{l=0}^{(1)}(r_0 = 0.01) = r_0$, $U_{l=0}^{\prime(1)}(r = r_0) \approx 1$.

The routine of the numerical integration of (2.24) has been given, e.g., in [2.2]. The conventional Milne method was used to obtain the solution itself. The integration was performed from $r = r_0$ up to $r = R = 20$, with $\Delta r = 0.08$ mesh width. Table 2.1 contains the results of the calculations. A comparison of these results with those obtained for the phase and cross section in the Born approximation shows that the deviation of the latter from the numerical solution is considerable. The comparison further shows that the S-phase shifts which give a dominant contribution to the cross section differ significantly. The difference between the P-phases, especially for $k > 0.2$ au, is, however, small. Thus, larger phases are described by the Born approximation quite satisfactorily.

The calculated scattering phase shifts and cross sections are also shown in Figs. 2.2, 3. It is seen that a small difference between the phases leads, however, to a considerable discrepancy between the cross sections. The curves in Fig. 2.3 give an idea of the accuracy of the Born approximation.

Table 2.1. Comparison of phase shifts [rad] and partial cross sections $|\pi a_o^2|$ for scattering of electrons in a potential given by $U = 2Zr^{-1}\exp(-r)$ calculated in the Born approximation (δ_i^B, σ_i^B) with those found by solving the equation (2.24)[a] (δ_i, σ_i)

k [au]	E [eV]	δ_0	δ_1	δ_0^B	δ_1^B	σ_0	σ_1	σ_0^B	σ_1^B
0.1	0.14	0.11	1.02(−3)	5.0(−2)	8.3(−5)	4.8	1.3(−3)	9.9(−1)	8.1(−6)
0.5	3.40	4.1(−1)	9.61(−3)	2.3(−1)	3.3(−3)	2.6	4.5(−3)	8.0(−1)	3.3(−3)
1.0	13.6	5.1(−1)	4.31(−2)	3.5(−1)	4.0(−2)	9.6(−1)	2.2(−2)	4.8(−1)	1.9(−2)
2.0	54.4	4.8(−1)	1.12(−1)	4.1(−1)	1.0(−1)	2.1(−1)	3.7(−2)	0.6(−1)	3.3(−2)
3.0	122.4	4.3(−1)	1.41(−1)	3.9(−1)	1.4(−1)	7.6(−2)	2.6(−2)	6.5(−2)	2.5(−2)
4.0	218	3.8(−1)	1.60(−1)	3.6(−1)	1.5(−1)	3.4(−2)	2.0(−2)	3.2(−2)	1.7(−2)

[a] The numbers in parentheses are the exponents on the power of 10

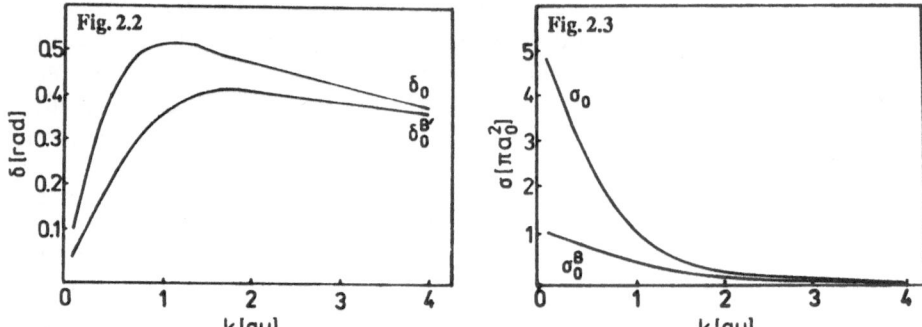

Fig. 2.2. Energy dependence of potential scattering phase shifts. The phase shift δ_0^B is obtained in the Born approximation, δ_0 is obtained by numerical solution of (2.24)

Fig. 2.3. Energy dependence of potential scattering cross sections. The cross section σ_0^B is obtained in the Born approximation, σ_0 is obtained by numerical solution of (2.24)

2.5 Resonances

Under certain conditions, a structure due to resonances can appear in the energy dependence of the potential scattering cross section. In this section, we present the theoretical interpretation of these resonances by considering the solution of the radial Schrödinger equation at complex energies. We shall confine ourselves to the case $l = 0$. A general solution of this equation in the asymptotic region is

$$U_0(r) = a(k)\exp(\mathrm{i}kr) + b(k)\exp(-\mathrm{i}kr) , \quad r \to \infty . \tag{2.38}$$

Here $a(k)$ and $b(k)$ are the particular cases of Jost functions [2.3]. It is possible to show that for many potentials, the Jost functions are analytic in the k-plane except for a cut along part of the imaginary axis [2.3–5]. We will use this assumption.

Consider, first of all, (2.38) for negative energies. Since $k^2 = 2E$, the pure imaginary wave number $k = \pm\mathrm{i}q(q > 0)$ corresponds to the negative $-q^2 = 2E$.

Then (2.38) assumes the form

$$U_0(r) = a(\mathrm{i}q)\exp(-qr) + b(\mathrm{i}q)\exp(qr) , \qquad \text{or}$$
$$U_0(r) = a(-\mathrm{i}q)\exp(qr) + b(-\mathrm{i}q)\exp(-qr) , \quad r \to \infty . \tag{2.39}$$

It follows that $a(\pm\mathrm{i}q) = b(\mp\mathrm{i}q)$. From the requirement that $U(r)$ be real and from (2.39) it is obvious that $a(\mathrm{i}q) = a^*(\mathrm{i}q)$. The solutions of (2.39) which describe the physical states must be bounded at infinity. This condition is true only if

$$b(\mathrm{i}q) = a(-\mathrm{i}q) = 0 . \tag{2.40}$$

Imposing the asymptotic condition (2.33c) written as

$$U_0(r) = \mathrm{const}\left[S_0\exp(-qr) + \exp(qr)\right] \tag{2.41}$$

and comparing (2.41) to (2.39) we have

$$S_0(\mathrm{i}q) = a(\mathrm{i}q)/b(\mathrm{i}q) . \tag{2.42}$$

It follows from (2.42) that the poles of the S-matrix correspond to the zeroes of the function $b(\mathrm{i}q)$.

Whereas the wave functions of the type given by (2.39) and the corresponding discrete negative energies describe bound states, it follows from (2.40, 42) that the S-matrix may have poles in the k-plane on the imaginary positive axis which correspond to these bound states.

Since at real k, according to (2.38), $b^*(k) = a(k)$ then

$$S_0^* S_0 = S_0 S_0^* = 1 . \tag{2.43}$$

Expression (2.43) states that the S-matrix is unitary. Note that this condition is valid only at real k. It turns out that the same analytic properties of the Jost function and S-matrix, as well as the same unitarity condition, hold for each l.

Recall the relation between the S-matrix and the scattering amplitude, see (2.26),

$$f_l = \left[\exp(2\mathrm{i}\delta_l) - 1\right]/2\mathrm{i}k = (S_l - 1)/2\mathrm{i}k . \tag{2.44}$$

Then

$$S_l S_l^* = 4k^2|f_l|^2 + 2\mathrm{i}k(f_l - f_l^*) + 1 = 1 ,$$

hence:

$$\mathrm{Im}\, f_l = k|f_l|^2 .$$

This expression is the optical theorem for the partial scattering amplitude. It follows from (2.44) that the poles of the scattering amplitude coincide with those of the S-matrix. These poles are positioned in the upper half-plane, and the zeroes in the lower one are conjugate to them. Indeed, (2.42) can be presented as

$$S(-iq) = a(-iq)/b(-iq) \tag{2.45}$$

and, since (2.40) holds, it is seen that the S-matrix has a zero at $k = -iq$ which is conjugate to the pole. Note that not any pole of the S-matrix corresponds to a bound state, but only those which are stipulated by the zeroes of the $b(iq)$-function.

It will be shown below that the presence of bound states with low binding energies manifests itself in a sharp increase of the cross section when the collision energy decreases to zero. However, cases are encountered when such an increase can be caused by the pole present in the lower half-plane which does not correspond to a bound state. According to (2.45), the S-matrix can have a pole located on the negative imaginary axis and stipulated by the zeroes of the $b(-iq)$-function. In this case, according to the second equation in (2.39), $U_0(r)$ is boundless as $r \to \infty$ and, thus, does not describe the bound state, but rather a virtual or nonbound state. Note that the true bound state must be located on the first sheet of the double-sheet energy surface, while the virtual one must be on the second sheet. This is true for nucleon-nucleon scattering, both in np- and pp-scattering. The cross section increases rapidly when the incident particle energy tends to zero. We shall return to the details of this question later.

Consider the singularities of the S-matrix when the energy is complex. Solving the time-dependent Schrödinger equation, we have

$$\Psi = \left[a \exp(ikr) + b \exp(-ikr) \right] \exp(-iEt) , \quad r \to \infty . \tag{2.46}$$

Considerations similar to those above lead us to the following result. The S-matrix, being a function of complex k, can have poles at $k = \lambda - i\kappa$, $k = -\lambda - i\kappa$ and zeroes at the symmetric points $k = -\lambda + i\kappa$, $k = \lambda + i\kappa$, but the occurrence of zeroes in the third and fourth quadrants (i.e., in the lower half-plane) is not allowed.

Since the solution of (2.46) at complex k is also restricted in r, i.e., it has the form (2.39), then it describes a certain bound state which is, however, damped with a time dependece $e^{-\gamma t}$, where $\gamma = \mathrm{Im}\, E$. This state is short-lived, with a lifetime

$$\tau \sim 1/\gamma . \tag{2.47}$$

Thus, we see that the S-matrix may be written [2.3]

$$S(k) = \prod_b \frac{k - k_b^*}{k - k_b} \prod_n \frac{(k - k_n^*)(k + k_n)}{(k - k_n)(k + k_n^*)} f(k) , \tag{2.48}$$

where k_b are poles located on the imaginary axis; k_n are those in the lower half-plane; $f(k)$ is a nonsingular function. It is easy to check that $S(k)S^*(k) = 1$.

How do the poles arise experimentally?

Consider first the case of pure imaginary poles. Let

$$S = -(k + iq)/(k - iq) , \tag{2.49}$$

where the minus sign is taken to provide the correct threshold behaviour of $S(0) = 1$. The partial S-wave cross section equals

$$\sigma_0 = \frac{4\pi}{k^2} |\sin \delta_0 \exp(i\delta_0)|^2 . \tag{2.50}$$

However, taking into account (2.45), one obtains

$$\sin \delta_0 \exp(i\delta_0) = \left[\exp(2i\delta_0) - 1\right]/2i = \frac{1}{2i}\left(-\frac{k + iq}{k - iq} - 1\right) .$$

Hence

$$\sigma_0 = 4\pi/(k^2 + q^2) . \tag{2.51}$$

If q is not very large, then as $k \to 0$, σ_0 may become rather large, i.e., the presence of a bound state manifests itself as mentioned above, in the sharp increase of the cross section with energy decrease. The bound states also correspond to $q > 0$. Negative q values are related to unbound or virtual states since according to (2.39), no bound state can exist at $q < 0$ because $U_0(r)$ is unlimited. In practice, however, this situation can indeed arise. It is obvious that negative imaginary k in this case testifies to the fact that the singularity on the two-sheeted Riemannian surface will be shifted to the second sheet. This singularity should not arise in the scattering amplitude explicitly.

It is known experimentally that, for instance, the total cross section for neutron-proton scattering (p + n → p + n) increases sharply when the energy decreases to zero. This is connected with the existence of a bound state, that is, a deuteron. However, a similar increase is observed in the pp-scattering cross section, in spite of the fact that Pauli's principle forbids the existence of a bound state in the p + p system. This is an example of a system with a virtual state. In electron-atom collisions, the increase of the cross section may result from the presence of a stable bound state, that is a negative ion of the target. For example, the growth of the e + H scattering cross ssection as $E \to 0$ is due to the possibility of formation of the $H^-(1s^2)$ negative ion.

Consider now the appearance of poles which correspond to the complex (but not pure imaginary) E values. Since there exists a pair of poles, the simplest possibility [2.3] is realized in accordance with (2.48)

$$S = (k + k_n)(k - k_n^*)/(k - k_n)(k + k_n^*) , \tag{2.52}$$

where $k_n = \pm\lambda - i\kappa$. The total cross section in this case is

$$\sigma_{tot} = \pi k^{-2}\Gamma^2/\left[(k^2 - k_0^2) + \Gamma^2/4\right] , \tag{2.53}$$

where

$$k_0^2 = \lambda^2 + \kappa^2 , \quad \Gamma = 4k\kappa . \tag{2.54}$$

Expression (2.53) is known as the Breit-Wigner formula. At $k = k_0$, σ_{tot} reaches its maximum, while Γ defines the resonance width. In the vicinity of k_0^2, σ_{tot} has a resonance form.

Hence, it follows that the complex poles of the S-matrix correspond to the resonances in the cross section and these resonances result from short-lived bound states, the lifetimes of which are given by (2.47).

Let us change the form of the expression for the resonant cross section. Let $S_l(k)$ have a single isolated pole near the real k-axis, the position of which is given by

$$k_r^2 = E_r - i\Gamma/2 . \tag{2.55}$$

Then at $E_r + i\Gamma/2$ the function $S_l(k)$ has a zero value, and we can represent approximately $S_l(k)$ in the vicinity of the pole by

$$S_l = e^{2i\delta_l^0}(E - E_r - i\Gamma/2)/(E - E_r + i\Gamma/2) , \tag{2.56}$$

where δ_l^0 is a slowly changing phase. The resonance phase shift is written as

$$\delta_l^{\mathrm{r}} = \arctan(\Gamma/2)(E - E_r)^{-1} , \tag{2.57}$$

so the total cross section is

$$
\begin{aligned}
\sigma_{tot} &= \frac{4\pi}{k^2} \sum_l (2l + 1) \sin^2\left(\delta_l^0 + \delta_l^{\mathrm{r}}\right) \\
&= \frac{4\pi}{k^2} \sum_l (2l + 1) \sin^2 \delta_l^0 \sin^2 \delta_l^{\mathrm{r}} \left(\cot \delta_l^0 + \cot \delta_l^{\mathrm{r}}\right)^2 \\
&= \frac{4\pi}{k^2} \sum_l (2l + 1)(q + \varepsilon)^2 (1 + \varepsilon^2)^{-1} \sin^2 \delta_l^0 .
\end{aligned}
\tag{2.58}
$$

Here $\varepsilon = -\cot \delta_l^{\mathrm{r}} = 2(E - E_r)/\Gamma$; $q = -\cot \delta_l^0$. Hence

$$\sigma_l = \sigma_l^0 (q + \varepsilon)^2 (1 + \varepsilon^2)^{-1} , \qquad \text{where}$$

$$\sigma_l^0 = 4\pi k^{-2}(2l + 1) \sin^2 \delta_l^0 . \tag{2.59}$$

This is the form for the resonant cross section often used in the literature.

The resonances must lie on the second Riemannian sheet of the E-plane. This means that the Breit-Wigner formula may be written in the form $f_{res} \sim \frac{1}{2}\Gamma/(E - E_r + i\Gamma/2)$, rather than $f_{res} \sim \frac{1}{2}\Gamma/(E - E_r - i\Gamma/2)$. This follows from the fact that the damping ot the time-dependent factor in (2.46) occurs only if $E = E_r - i\Gamma/2$. However, the values of E mentioned above are realized only if the singularities are located in the lower half-plane of k. Note that at the resonance point, $\sigma_{res} = \sigma_{tot} - \sigma_l^0$ reaches the geometric value $\sigma_{res} = 4\pi/k^2$. Thus, only those peaks in the total elastic cross section where σ_{res} reaches the geometric cross section values may be called the true resonances.

Consider the conditions under which a resonance in potential scattering occurs. In atomic physics resonances of this type are called shape resonances. We use the "effective-range approximation" [2.1, 3]. Consider again, for simplicity, an S-wave, i.e., let $l = 0$. Within this approximation, $\cot \delta_0$ can be presented for small k as

$$k \cot \delta_0 = -1/a + r_0 k^2/2 + \ldots . \tag{2.60}$$

The value of a thus chosen is called the scattering length and r_0 is called the effective range. It is known that the scattering amplitude can be written in the form

$$f_0 = k^{-1}/(\cot \delta_0 - i) = \left(-1/a + r_0 k^2/2 - ik\right)^{-1} . \tag{2.61}$$

Denoting $ik = \kappa$, we find the roots of the equation $-r_0 \kappa^2/2 - \kappa - 1/a = 0$. Then $\kappa_{1,2} = -1/r_0 \pm r_0^{-1}(1 - 2r_0/a)^{1/2}$. If $2r_0/a < 1$, then κ is real, k is pure imaginary, and a and r_0 correspond to bound states.

Take the case when $2r_0/a > 1$. It appears that $k_{1,2} = i/r_0 \pm r_0^{-1}(2r_0/a-1)^{1/2}$, and (2.61), as can be seen from the direct calculation, leads to a resonance in the cross section if a and r_0 are negative. This can occur for some specific forms of the interaction potential, for example, those with a rather deep well and a barrier (Fig. 2.4). It appears that no realistic atomic potential exists which may generate that resonance for $l = 0$.

Sometimes in the literature the increase of the cross section as $k \rightarrow 0$ is also called, according to (2.51), the resonance at $l = 0$. However, we know that actually this pattern is caused by the bound (or virtual) state. The true resonance can arise only if $l \neq 0$, since the presence of a centrifugal term $l(l + 1)/r^2$ is necessary. The sum of certain short-range attractive potentials and $l(l + 1)/r^2$ can have the form shown in Fig. 2.4. Thus it is said that shape resonances result from a centrifugal barrier. When such a barrier is present, the bound states of the attractive potential (dotted line in Fig. 2.4) appear in the continuous spectrum as a shape resonance. This happens usually at energies which slightly exceed the threshold of the process, and, therefore, the width of the shape resonance is relatively large.

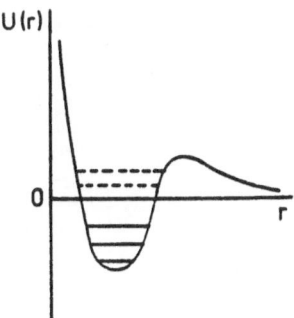

Fig. 2.4. A potential which leads to shape resonances. (—) bound states, (- - -) resonance states

Figure 7.49 shows a realistic effective potential $U = V_{\text{atom}} + l(l+1)/r^2$ which can be used to describe the elastic e + Ca scattering (Chap. 7). This potential causes a shape resonance in the D-wave ($l = 2$). Shape resonances are well known in elastic electron scattering by some alkali (Li, Na, K) and alkaline-earth (Mg, Ca, Sr, Ba) atoms. There are some indications of the occurrence of shape resonances in excitation cross sections as well. This problem will be discussed in detail in the following chapters.

Note that the effective-range approximation, as has been previously shown [2.3], is valid for short-range potentials. For long-range potentials (2.60) must be modified. For example, if the potential $U(r)$ behaves like α/r^4, then for $l = 0$ [2.6]

$$k \cot \delta_0 = -1/a + \pi\alpha k/3a^2 + (3a)^{-1}2\alpha k^2 \ln(\alpha k^2/16) + O(k^2) . \qquad (2.62)$$

At low $k \to 0$ (2.62) can be rewritten in the form

$$\tan \delta_0 \approx -ak - \pi\alpha k^2/3 . \qquad (2.63)$$

Hence it follows that if $a < 0$, the phase shift first increases with increasing k, then decreases, and may vanish at $k_0 = -3a/\pi\alpha$. Since the S-phase is dominant at low k, this means that the cross section has a deep minimum in this region. This phenomenon was indeed observed in electron scattering by inert gas atoms and was called the Ramsauer-Townsend effect. In this case the long-range attractive potential is a polarization potential which results from the deformation of the valence atomic shells by the incident electron.

It should be noted that a generalization of the results concerning analytical properties of the amplitude of electron-atom scattering has not yet been accomplished [2.7].

3. Collisions Between an Electron and a Composite Target

The theory discussed in the previous chapter is applied here to describe scattering from composite targets. We study the elastic and inelastic electron scattering by a hydrogen atom without closed channels being taken into account.

Theoretical methods used widely to describe the scattering processes, such as the static approximation, distorted-wave approximation and close-coupling method are described using simple examples. The noniterative procedure of solution of the integro-differential equations arising in the case when exchange is taken into account is considered. This chapter is transitional, on the way from potential scattering to the theory of resonances.

3.1 Static Potential Approximation

Electron scattering by atomic hydrogen (or a hydrogen-like ion with nuclear charge Z) in the nonrelativistic approximation is described by the Schrödinger equation (without spin-orbit interaction)

$$H\Psi = E\Psi , \tag{3.1}$$

where E is the total energy of the system and

$$H = -\frac{1}{2}\nabla_1^2 - \frac{1}{2}\nabla_2^2 - \frac{Z}{r_1} - \frac{Z}{r_2} + \frac{1}{r_{12}} \tag{3.2}$$

is the Hamiltonian of the system.

Let us approximate the solution of (3.1) by the function Ψ taken in the separable form

$$\Psi(r_1, r_2) = F(r_1)\phi(r_2) . \tag{3.3}$$

This choice of the Ψ-function corresponds to the single-particle approximation. Here we consider only the coordinate part of the wave function and do not take into account the exchange; this will be done in the following section. Here $\phi(r_2)$ is the wave function of the groundstate of a hydrogen atom ($n = 1, l = 0, m = 0$) with the eigenvalue E_1, i.e.,

$$\left(-\frac{1}{2}\nabla_2^2 - \frac{Z}{r_2}\right)\phi(r_2) = E_1\phi(r_2) . \tag{3.4}$$

This assumption implies that the incident electron in no way affects the target state, i.e., the target is "frozen". This assumption is called the static approximation.

Substituting (3.3) into (3.1) and taking (3.4) into account we may write

$$\left(-\frac{1}{2}\nabla_1^2 + E_1 - \frac{Z}{r_1} + \frac{1}{r_{12}}\right) F(\mathbf{r}_1)\phi(\mathbf{r}_2) = EF(\mathbf{r}_1)\phi(\mathbf{r}_2) \ . \tag{3.5}$$

Let us multiply (3.5) by $\phi^*(\mathbf{r}_2)$ and integrate the product over $\mathbf{r}_2$. Keeping in mind the orthogonality of hydrogen wave functions, we obtain the following equation:

$$\left[\nabla_1^2 + k_1^2 - 2V(\mathbf{r}_1)\right] F(\mathbf{r}_1) = 0 \ , \tag{3.6}$$

where

$$V(\mathbf{r}_1) = \int d\mathbf{r}_2 |\phi(r_2)|^2 \left(\frac{1}{r_{12}} - \frac{Z}{r_1}\right) \ ; \quad E = \frac{k_1^2}{2} + E_1 \ , \tag{3.7}$$

and k_1 is the momentum of the incident electron. $V(\mathbf{r}_1)$ is called a static potential for the hydrogen atom.

It should be mentioned that in fact (3.5), which is considered to be an exact equation for a many-particle problem, has no solution. In reality, (3.6) is obtained from the condition $\langle\phi|H - E|\Psi\rangle = 0$, which follows from (3.1). Note that this condition is necessary but not sufficient for Ψ to be a solution of (3.1) However, such a Ψ approximates, in the best manner, the exact solution in the "Bubnov-Galerkin sense" [3.1]. The variational principle is one of the alternative ways of providing the equation for $F(\mathbf{r})$.

Let us calculate the static potential. For the ground state of the hydrogen atom (or a hydrogen-like ion with charge Z)

$$\phi_{100}(\mathbf{r}) = (\pi)^{-1/2} Z^{3/2} e^{-Zr} \ . \tag{3.8}$$

Consider the two terms in $V(\mathbf{r}_1)$ separately. Using the expansion

$$\frac{1}{r_{12}} = \sum_{l=0}^{\infty} \frac{r_<^l}{r_>^{l+1}} P_l(\cos\Theta) \tag{3.9}$$

to calculate the first term in $V(\mathbf{r}_1)$, we obtain

$$V^{(1)} = 4Z^3 \left[-(1/4Z^3 r_1 + 1/4Z^2)e^{-2Zr_1} + 1/4Z^3 r_1\right] \ . \tag{3.10}$$

Here $r_<$ and $r_>$ are the smallest and largest values from a pair (r_1, r_2), respectively.

Due to the normalization of the $\phi(\mathbf{r}_2)$-function, the integration of the second term in (3.7) produces

$$V^{(2)} = -\frac{1}{r_1} Z \ . \tag{3.11}$$

Adding both parts we obtain the equality:

$$V(r_1) = -e^{-2Zr_1}(1 + 1/Zr_1)Z - (Z - 1)/r_1 \ . \tag{3.12}$$

It was noted above that this result is of particular significance. The long-range part of the interaction between the incident and atomic electrons and the attraction of the incident electron by the hydrogen atom nucleus compensate for each other exactly. Only the short-range part of the interaction survives, being of an attractive character [which is indicated by the minus sign in (3.12)]. It turns out that the attraction of a nucleus plays a more significant role although the nucleus is screened by the atomic electron. An interaction of this type is called a screened interaction. Electron scattering by a hydrogen atom within the static approximation occurs in the same manner as in the potential field.

Let us emphasize that the static potential field of the hydrogen atom is a short-range field which does not allow description of the important details of the scattering. In order to obtain more realistic results one has to take into account the influence of the incident electron on the target, i.e., the possibility of the target to be changed. In the following chapters, we shall consider the more general case when the electron can be captured by the excited target in the field of the potential which is realized instead of the potential (3.12) with the subsequent formation of an AIS. This process results just in the production of the resonances in the scattering.

It should be noted that the first term in (3.12) coincides qualitatively with the Yukawa potential for strong interactions. The interaction of, say, a proton with a π-meson at large distances, obeys the same law. This suggests that the Yukawa potential is, in reality, the screened one-gluon potential which, in the nonrelativistic approximation, has the form $V_{one-gluon} \sim e^{-\mu r}/r$. This fact, in any case, testifies once more to the complex (quark) structure of the elementary particles.

3.2 Radial Equation for Elastic Scattering

It is quite a complicated problem to obtain an analytic solution of (3.1) within the perturbation method [3.2]. The Born approximation overestimates the cross section in the low energy region. It is quite a labourious and cumbersome task to obtain results within the framework of subsequent approximations. Therefore, it is necessary to look for other methods to describe the scattering. The direct numerical integration of the equations of motion on a computer is one method which gives a satisfactory result. It is based on radial equations since there are still no algorithms of solution of three-dimensional equations. So, let us obtain from (3.6) an equation for a radial function. It is known that the solution of such equations can be represented in the following form [similarly to (2.22)]:

$$F(r) = \sum_l [4\pi(2l + 1)]^{1/2} i^l \exp(i\delta_l) \frac{f_{lm}(r)}{k_1 r} Y_{lm}(\Omega) \ , \tag{3.13}$$

where

$$\Omega = (\Theta, \phi) .$$

A few words should be said about the expansion (3.13) of the solution of the Schrödinger equation over partial waves and about its difference from the expansion (2.22) for the potential scattering. The difference lies in the dependence of the wave function on the azimuthal angle ϕ (in the general case). Thus in (3.13) the spherical functions $Y_{LM}(\Omega)$ are present instead of the Legendre polynomials $P_L(\cos \Theta)$.

Generally speaking, in (3.13) summation must be performed over m too. However, due to the conservation of orbital angular momentum projection:

$$m + m_1 = M , \tag{3.14}$$

(where m_1 is the projection of the atomic electron orbital angular momentum, M is that of the total orbital angular momentum), summation is, actually, not realized. This is a result of the invariance of the Hamiltonian (3.2) under spatial rotations.

Then, substituting (3.13) into (3.5) and using the properties of the Laplacian in the spherical coordinate frame, we get

$$\sum_l (4\pi(2l + 1))^{1/2} i^l \left(\exp(i\delta_l)/k_1 r\right) Y_{lm}(\Omega)$$

$$\times \left[f''_{lm}(r) + \left(k_1^2 - \frac{l(l+1)}{r^2} - 2V(r) \right) f_{lm}(r) \right] = 0 . \tag{3.15}$$

Let us multiply (3.15) by $Y^*_{l'm'}(\Omega)$ and use the orthogonality condition of the spherical functions. Using (3.12) at $Z = 1$ and taking into account (2.17), we obtain the radial equation

$$\left[\frac{d^2}{dr^2} + k_1^2 - \frac{l(l+1)}{r^2} - U(r) \right] f_{lm}(r) = 0 , \quad U = 2V . \tag{3.16}$$

Equation (3.16) coincides with (2.24) obtained in Chap. 2. This results from the condition that within the static approximation the potential is spherically symmetric, like the potential in (2.24). The latter equation should be solved with boundary conditions (2.26). In the general case it may be solved, e.g., by such numerical methods as Milne's method, Numerov's method or the implicit alternation-direction method. The calculated phase shifts are presented in Table 3.1. Since the static potential is known not to lead to the formation of a bound state, then $\delta_0 \rightarrow 0$ as $k_1 \rightarrow 0$. Let us compare the results from Table 3.1 with those obtained in the Born approximation. It appears that at $k_1 = 1$ au the S-phases differ by a factor of 1.5; while the cross sections differ by a factor of 2. However, at $k_1 = 2$ au the Born phase is equal to $\delta_0^B = 0.627$ rad which is close to the value presented in Table 3.1. Hence, at this energy ($E \approx 55$ eV) the Born approximation becomes valid.

Table 3.1. Phase shifts [rad] for the elastic scattering of electrons by an H atom found in the static approximation

k [au]	δ_0	δ_1	δ_2	k [au]	δ_0	δ_1	δ_2
0.2	0.97	0.00	0.00	2.0	0.70	0.22	0.02
0.5	1.06	0.03	0.00	3.0	0.57	0.25	0.12
1.0	0.91	0.11	0.08	4.0	0.49	0.26	0.17

All previous results, as was mentioned above, were obtained without exchange. According to the Pauli principle, the total wave function of the system must be antisymmetric with respect to the transposition of two electrons. The reasons are known, and lead to the following expression for the spatial part of the wave function in a symmetric form within the static approximation [3.3]

$$\Psi^\pm(r_1, r_2) = \frac{1}{\sqrt{2}} \left[F^\pm(r_1)\phi(r_2) \pm F^\pm(r_2)\phi(r_1) \right] . \tag{3.17}$$

The sign "$\pm$" stands for a singlet (triplet) state, respectively. The Pauli principle is reflected in (3.17): when the particles are transposed, their spatial wave function $\Psi^\pm(r_1, r_2)$ changes its sign in the case of a triplet state and retains it for a singlet state.

Let us substitute (3.17) into (3.1) where the Hamiltonian H is given by (3.2). Note that $\phi(r)$ obeys (3.4). Multiplying both sides of the obtained expression by $\phi^*(r_2)$ and integrating it over r_2 we obtain, taking into account the orthogonality of the ϕ-functions

$$\nabla_1^2 F^\pm(r_1) + k_1^2 F^\pm(r_1) = F^\pm(r_1)U(r_1) \pm \phi(r_1)\left[\int \phi^*(r_2)\right.$$
$$\left. \times \left(-\nabla_2^2 - k_1^2\right) F^\pm(r_2)dr_2 + \int \phi^*(r_2)U_2(r_1, r_2) F^\pm(r_2)dr_2\right] , \tag{3.18}$$

where k_1 and E are related by (3.7) and $U_2 = 2(1/r_{12} - 1/r_2)$.

Using the Hermiticity of the operator ∇_2^2

$$\int \phi_n^*(r_2)\nabla_2^2 F_m(r_2)dr_2 = \int F_m(r_2)\nabla_2^2\phi_n^*(r_2)dr_2 ,$$

and keeping in mind that $\phi(r_2)$ obeys (3.4), we may write the following expression:

$$\int dr_2\phi^*(r_2)\nabla_2^2 F^\pm(r_2) = -2 \int dr_2 \left(E_1 + \frac{1}{r_2}\right) F^\pm(r_2)\phi^*(r_2) .$$

Introducing the notation

$$W(r_1, r_2) = 2\phi(r_1)\phi^*(r_2)\left(E_1 - \frac{k_1^2}{2} + \frac{1}{r_{12}}\right) , \tag{3.19}$$

23

we may write (3.18) in the form

$$\nabla_1^2 F^{\pm}(\mathbf{r}_1) + k_1^2 F^{\pm}(\mathbf{r}_1)$$

$$= U(r_1)F^{\pm}(\mathbf{r}_1) \pm \int W(\mathbf{r}_1, \mathbf{r}_2)F^{\pm}(\mathbf{r}_2)d\mathbf{r}_2 . \tag{3.20}$$

It is seen that the right-hand side of (3.18) and (3.20) are different. The possible influence of this fact on further results will be discussed in the following section.

The radial equation for the symmetrized wave function can be obtained from (3.20). To do this, let us introduce the following notation: l_1, m_1 denote an atomic electron; while l_2, m_2 belong to an incident electron; l, m are the total momenta of the system.

If we substitute (3.13) into (3.20), multiply both sides of the obtained equation by $Y_{lm}^*(\Omega_1)$, and then integrate it over Ω_1, we get the following equation:

$$\left[\frac{d^2}{dr_1^2} - \frac{l(l+1)}{r_1^2} + k_1^2 - V(r_1) \right] f_{lm}^{\pm}(r_1)$$

$$= \pm \sum_{l'} \left(\frac{2l'+1}{2l+1} \right)^{1/2} e^{i(\delta_l - \delta_{l'})} i^{l-l'} \int W_{lml'm}(r_1, r_2)$$

$$\times f_{l'm}^{\pm}(r_2) r_1 r_2 dr_2 , \tag{3.21}$$

where the new potentials V and W are, however, functions of the moduli of r

$$V(r_1) = \sum_{l'} \int Y_{l'm}^*(\Omega_1)U(r_1)Y_{lm}(\Omega_1)d\Omega_1 , \tag{3.22}$$

$$W_{lml'm'}(r_1, r_2) = \int Y_{l'm'}^*(\Omega_1)W(\mathbf{r}_1, \mathbf{r}_2)Y_{lm}(\Omega_2)d\Omega_1 d\Omega_2 . \tag{3.23}$$

The potential $W_{lml'm'}$ is called the exchange potential. The expressions for $U(r_1)$ and $W(\mathbf{r}_1, \mathbf{r}_2)$ are already calculated [(2.17), (3.12, 19)]. The calculation of $V(r_1)$ is not difficult since $U(r_1)$ depends only on the modulus of r_1.

It follows from (3.23) that due to the orthogonality of the spherical functions, $V(r_1) = U(r_1)$. The expression for the exchange potential can be simplified, according to (3.19) and (3.23), for the elastic scattering by a hydrogen atom by using the expansion of r_{12} over the spherical functions [3.4]

$$\frac{1}{r_{12}} \equiv \frac{1}{|\mathbf{r}_1 - \mathbf{r}_2|} = \sum_k \frac{4\pi}{2k+1} \gamma_k \sum_{q=-k}^{k} Y_{kq}(\Omega_1)Y_{kq}^*(\Omega_2) . \tag{3.24}$$

Here $\gamma_k = r_<^k / r_>^{k+1}$. Then, using the orthogonality condition, we have

$$W_{lml'm'} = 2R_{10}(r_1)R_{10}(r_2) \left[\frac{r^l}{r^{l+1}} + \left(E_1 - \frac{k_1^2}{2} \right) \delta_{l0}\delta_{m0} \right] \delta_{ll'}\delta_{mm'} , \tag{3.25}$$

where $R_{10}(r) = 2\exp(-r)$ is the radial function of the ground state. Thus, we obtain the final result as follows:

$$\left[\frac{d^2}{dr_1^2} + k_1^2 - \frac{l(l+1)}{r_1^2}\right] f_{lm}^{\pm}(r_1)$$

$$= V(r_1)f_{lm}^{\pm}(r_1) \pm \int dr_2 r_1 r_2 W_{lmlm}(r_1, r_2)f_{lm}^{\pm}(r_2) . \qquad (3.26)$$

Let's say a few words about the solution of (3.26). Due to the exchange this equation turns out to be integro-differential. This equation can be solved in different ways. The most successful one was proposed by it Drukarev as far back as in 1953. We shall follow [3.5] expounding his method.

For simplicity, we shall analyze the elastic *s*-scattering ($l = 0$) of an electron by a hydrogen atom. Thus

$$\left[\frac{d^2}{dr_1^2} + k_1^2 - V(r_1)\right] f_{00}^{\pm}(r_1) = \pm \int_0^{\infty} dr_2 r_1 r_2 W_{0000}(r_1, r_2)f_{00}^{\pm}(r_2) , \qquad (3.27)$$

where

$$W_{0000}(r_1, r_2) = \begin{cases} W_{00}^{(1)}(r_1, r_2) \\ W_{00}^{(2)}(r_1, r_2) \end{cases}$$

$$= \begin{cases} 2R_{10}(r_1)R_{10}(r_2)\left(1/r_1 + E_1 - k_1^2/2\right) , & r_1 > r_2 \\ 2R_{10}(r_1)R_{10}(r_2)\left(1/r_2 + E_1 - k_1^2/2\right) , & r_1 < r_2 . \end{cases} \qquad (3.28)$$

Now the integral [see (3.26)] may be written as

$$\int_0^{r_1} \left(W_{00}^{(1)} - W_{00}^{(2)}\right) r_1 r_2 f_{00}^{\pm}(r_2)dr_2 + 2P^{\pm} r_1 R_{10}(r_1) , \qquad (3.29)$$

where $P^{\pm}$ is a number:

$$P^{\pm} = \int_0^{\infty} (1/r_2 + E_1 - k_1^2/2)r_2 R_{10}(r_2) f_{00}^{\pm}(r_2)dr_2 .$$

Let us now write the radial equation (3.27) using the explicit form of $R_{10}(r)$

$$\left[\frac{d^2}{dr_1^2} + k_1^2 - V(r_1)\right] f_{00}^{\pm}(r_1)$$

$$= \pm 8r_1 e^{-r_1} \int_0^{r_1} r_2 dr_2 f_{00}^{\pm}(r_2)(1/r_1 - 1/r_2)e^{-r_2} \pm 4e^{-r_1} P^{\pm} . \qquad (3.30)$$

Introducing the functions:

$$f_1^{\pm}(r_1) = \int_0^{r_1} r_2 dr_2 e^{-r_2} f_{00}^{\pm}(r_2) ,$$

$$f_2^{\pm}(r_1) = \int_0^{r_1} dr_2 e^{-r_2} f_{00}^{\pm}(r_2) , \qquad (3.31)$$

and the constant:

$$f_3^\pm = \int_0^\infty B f_{00}^\pm(r_2) dr_2 \,, \quad B = e^{-r_2}\left(\frac{1}{r_2} + E_1 - \frac{k_1^2}{2}\right) r_2 \,, \tag{3.32}$$

we rewrite (3.30) as

$$\left[\frac{d^2}{dr_1^2} + k_1^2 - V(r_1)\right] f_{00}^\pm(r_1)$$
$$= \pm 8 e^{-r_1} f_1^\pm(r_1) \mp 8 e^{-r_1} r_1 f_2^\pm(r_1) \pm 8 e^{-r_1} r_1 f_3^\pm \,. \tag{3.33}$$

If one supplements (3.33) with the equations implied by (3.31)

$$\frac{df_1^\pm(r_1)}{dr_1} = r_1 e^{-r_1} f_{00}^\pm(r_1) \,; \quad \frac{df_2^\pm(r_1)}{dr_1} = e^{-r_1} f_{00}^\pm(r_1) \,, \tag{3.34}$$

then (3.34) together with (3.33) form a system of inhomogeneous equations with the inhomogeneous term $\pm 8 r_1 e^{-r_1} f_3^\pm$. The corresponding solutions of the homogeneous system are denoted here as $F_{1,2}^{\pm(0)}$. Then

$$F^{\pm(0)} = C_1 F_1^{\pm(0)} + C_2 F_2^{\pm(0)} \,. \tag{3.35}$$

Here F_1 is a regular, while F_2 is a nonregular solution. Assume on the basis of physical considerations that $C_2 = 0$ (similarly as in Sect. 2.4). We shall denote the partial solution of the inhomogeneous system as $F^{\pm(i)}(r)$; then

$$\left[\frac{d^2}{dr_1^2} + k_1^2 - V(r_1)\right] F^{\pm(i)}(r_1)$$
$$= \pm 8 e^{-r_1} f_1^{\pm(i)}(r_1) \mp 8 e^{-r_1} r_1 f_2^{\pm(i)}(r_1) \pm 8 e^{-r_1} r_1 f_3^\pm \,;$$
$$\frac{df_1^{\pm(i)}(r_1)}{dr_1} = r_1 e^{-r_1} F^{\pm(i)}(r_1) \,;$$
$$\frac{df_2^{\pm(i)}(r_1)}{dr_1} = e^{-r_1} F^{\pm(i)}(r_1) \,. \tag{3.36}$$

The general solution of this system is a sum of the partial solution of the inhomogeneous equation and the general solution of the homogeneous equation, i.e.,

$$f_{00}^\pm(r_1) = C_1 F_1^{\pm(0)}(r_1) + f_3^\pm F^{\pm(i)}(r_1) \,. \tag{3.37}$$

It is easy to be convinced that (3.37) is indeed the solution of (3.33). To do that, one should insert (3.27) into (3.33) and take into account that $F_1^{\pm(0)}$ is a solution of the corresponding homogeneous equation and $F^{\pm(i)}$ is that of the inhomogeneous one, (3.36). It must be kept in mind that $f^\pm = f^{\pm(0)} + f_3^\pm f^{\pm(i)}$.

The unknown constant $f_3^\pm$ is determined by the substitution of (3.37) into (3.32)

$$f_3^\pm = \int_0^\infty B \left[C_1 F_1^{\pm(0)}(r_1) + f_3^\pm F^{\pm(i)}(r_1) \right] dr_1 \,. \tag{3.38}$$

Thus, we have obtained some algebraic equation for $f_3^{\pm}$, the solution of which has the following form:

$$f_3^{\pm} = \frac{C_1 \int_0^\infty BF_1^{\pm(0)}(r_1)dr_1}{1 - \int_0^\infty BF^{\pm(i)}(r_1)dr_1} . \tag{3.39}$$

The solution of (3.30) may be written as

$$f_{00}^{\pm}(r_1) = C_1 \left[F_1^{\pm(0)}(r_1) + Y(r_1) \right] ; \quad Y(r) = \frac{f_3^{\pm}}{C_1} F^{\pm(i)}(r) . \tag{3.40}$$

So the problem of finding the solution of the integro-differential equation (3.27) is reduced to a more general problem of finding the general solution of the system (3.33, 34) of homogeneous equations and the partial solution of the system (3.36) of inhomogeneous equations. The routines for finding the numerical solutions of these systems are quite long, and we shall omit them since they may be found elsewhere [3.2].

The constant C_1 and the scattering amplitude are determined from the condition of matching the solution of (3.40) with the asymptotics ($K \equiv K_1$)

$$C_1 \left[F_1^{\pm(0)}(R) + Y(R) \right] = \frac{1}{k} \sin kR + K_0^{\pm} \cos kR ;$$

$$C_1 \left[F_1'^{\pm(0)}(R) + Y'(R) \right] = \cos kR - kK_0^{\pm} \sin kR , \tag{3.41}$$

where $K_0^{\pm} = \tan \delta_0^{\pm}$; $\delta_0^{\pm}$ are the singlet and triplet scattering phases. It is easy to determine from (3.41) that

$$K_0^{\pm} = \frac{A \cos kR - (1/k)) A' \sin kR}{A' \cos kR + kA \sin kR} ,$$

where $A = F_1^{\pm(0)} + Y(R)$. The numerical solution of the integro-differential equation (3.26) gives the phase shift values presented in Table 3.2. Note that $\delta_0^{\pm} \to \pi$ as $k \to 0$. This corresponds to Levinson's theorem which states that the difference between the phase shifts at a zero energy and at an infinite energy equals the number of bound and antibound states multiplied by π.

As has been pointed out in a series of studies [3.6], the interpretation of the atom as a scattering center described by a static potential, which implies that

Table 3.2. Singlet (δ_0^+) and triplet (δ_0^-) S-phase shifts for the elastic scattering of electrons by an H atom found in the static approximation with exchange

k [au]	δ_0^+	δ_0^-	k [au]	δ_0^+	δ_0^-
0.0	8.95[a]	2.35[a]	0.6	0.87	1.90
0.2	1.87	2.68	0.8	0.65	1.61
0.4	1.24	2.26	1.0	0.54	1.39

[a] Scattering length

only the initial state of the atom is considered, leads to incorrect results. This results from the distortion of the atomic field under the approach of an incident electron. This distortion effect can be interpreted as a virtual excitation of higher levels which results in the appearance of an additional polarization potential

$$V_{\text{pol}}(r_1) = \sum_{\alpha \neq 1} \frac{|\langle \phi_\alpha|V|\phi_1\rangle|^2}{E_\alpha - E_1} \,, \tag{3.42}$$

where

$$\langle \phi_\alpha|V|\phi_1\rangle = \int \phi_\alpha^*(r_2)V(r_1, r_2)\phi_1(r_2)dr_2 \,;$$
$$V(r_1, r_2) = -1/r_1 + 1/r_{12}$$

is the interaction potential for the incident electron [see (3.2)]; ϕ_α describes the excited states of the target.

It appears that $V_{\text{pol}}(r)$ decreases as $(-\alpha/2r^4)$ at large r where

$$\alpha = \sum_{\alpha \neq 1} \frac{|\langle \phi_\alpha|r|\phi_1\rangle|^2}{E_\alpha - E_1} \,, \tag{3.43}$$

i.e., α has the form of a dipole matrix element.

One can show that choosing the system of wave functions in the form

$$\Psi(r_1, r_2) = \sum_\alpha F_\alpha(r_1)\phi_\alpha(r_2) + \sum_\beta \int F_\beta(E, r_1)\phi_\beta(E, r_2)dE \tag{3.44}$$

where the summation over α runs over all discrete atomic states and the second term describes the electron in a continuous spectrum, upon the substitution of (3.44) into (3.1) [3.3] we arrive at terms similar to (3.42). The total polarizability for a hydrogen atom is $\alpha = 4.5$ au. The sum (3.43) over discrete states contributes 3.7 au to this value while all the remaining contribution comes from the continuous spectrum. The first term in (3.43) provides $\alpha = 2.96$ au, hence, the sum (3.44) can be restricted to a few terms.

Sometimes one can use phenomenologically more complex polarized potentials, e.g.,

$$V_{\text{pol}} = \frac{\alpha(r)}{2r^4} \,, \qquad \text{where}$$
$$\alpha(r) = \tfrac{9}{2}\left[1 - e^{-(r/\varrho)^6}\right] \,; \qquad \text{(in au)} \,, \tag{3.45}$$

and ϱ is a fitting parameter.

It is necessary to introduce the potential (3.45) because the phenomenological potential diverges at zero too strongly. The potential (3.45) obeys the condition $V \sim r^2$ at small r. One may show, in fact, that $V_{\text{pol}}(r \to 0) \to$ const [3.6] due to the monopole polarization of the atom, which corresponds to some "breathing modes" originating from a symmetric expansion (compression) of the target.

It is seen, however, from (3.45) that the phenomenological specification of the potential requires the introduction of an increasing number of fitting parameters. Therefore we shall give up this method. Another approach to this problem, namely, the consideration of the possibility of a change in the target state is detailed below.

As follows from Table 3.2, the singlet and triplet phase shifts differ considerably at low energies. This fact reveals the significance of the exchange. The influence of the latter is reduced with increasing energy, and the phase shifts tend to the "exchangeless" values presented in Table 3.1.

3.3 Inelastic Scattering.
The Distorted-Wave Approximation

Consider first of all the simplest case of excitation, where the nonsymmetrized wave function is taken, i.e., neglect the exchange. The differential excitation cross section for the α-th state of a hydrogen atom can be expressed in terms of the scattering amplitude $f(k, \theta)$

$$\frac{d\sigma_\alpha(k_0, \theta)}{d\Omega} = \frac{k_\alpha}{k_0} |f_\alpha(k_0, \theta)|^2 , \tag{3.46a}$$

as well as the total cross section

$$\sigma_\alpha(k_0) = \frac{k_\alpha}{k_0} \iint |f_\alpha(k_0, \theta)|^2 \sin\theta \, d\theta \, d\phi . \tag{3.46b}$$

Here k_0 and k_α are electron momenta in the elastic and excitation channels, respectively, $E = E_\alpha + k_\alpha^2/2$.

Let us start with the Schrödinger equation (3.1) where H is the total Hamiltonian of the system (3.2). We shall look for the solution in the form of (3.44) by means of generalization of (3.3) to the case of an inelastic process. Here $\phi_\alpha(r)$ is the wave function of a hydrogen atom in the α-th state; α is a generalized quantum number, $\alpha \equiv (n, l_1, m_1)$, which characterizes the state of the atom.

Having the test function as in (3.44), we have made a significant step forward: we have taken into account a possible perturbation of the target state by the incident electron, namely, a possible transition into a certain excited state (real or virtual). Inclusion of all possible terms in (3.44) is sufficient for a complete consideration of the polarization. In practice, it is, of course, impossible to take into account an infinite number of excited states. Usually, only a few terms in (3.44) are retained. The second term corresponds to the transition of the atomic electron into a continuous spectrum. In cases where the ionization is beyond consideration, the second term usually is neglected, which is, of course, an additional simplifying assumption. Complete consistent inclusion of this term is far beyond our present capabilities. We shall discard this term in the further analysis.

Upon substituting the solution of (3.44) into the Schrödinger equation and taking into account that $\phi_\alpha(\mathbf{r}_1)$ satisfies an equation of the type of (3.4) we get

$$\sum_\alpha \phi_\alpha(\mathbf{r}_1)\left[\nabla_2^2 + k_\alpha^2 + 2\left(\frac{1}{r_2} - \frac{1}{r_{12}}\right)\right]F_\alpha(\mathbf{r}_2) = 0 . \tag{3.47}$$

Basing ourselves on the remarks in the derivation of (3.6), we obtain the equation for $F_\alpha(\mathbf{r}_2)$

$$\left(\nabla_2^2 + k_\alpha^2\right)F_\alpha(\mathbf{r}_2) = 2\sum_\beta F_\beta(\mathbf{r}_2)V_{\beta\alpha}(\mathbf{r}_2) , \tag{3.48}$$

where

$$V_{\beta\alpha}(\mathbf{r}_2) = \int \phi_\beta^*(\mathbf{r}_1)V_2(\mathbf{r}_1,\mathbf{r}_2)\phi_\alpha(\mathbf{r}_1)d\mathbf{r}_1 ,$$

$$V_2(\mathbf{r}_1,\mathbf{r}_2) = -1/r_2 + 1/r_{12} .$$

Let us select those solutions $F_\alpha(\mathbf{r}_2)$ for which (as $r \to \infty$)

$$F_\alpha(\mathbf{r}) \sim e^{ik_\alpha r}\delta_{\alpha 0} + f_\alpha(k_0, \theta)\frac{e^{ik_\alpha r}}{r} . \tag{3.49}$$

The solution of (3.47) may be found by means of the Green's function $G(\mathbf{r}_2, \mathbf{r}_2')$

$$F_\alpha(\mathbf{r}_2) = e^{ik_0 \mathbf{r}_2}\delta_{\alpha 0} + 2\int d\mathbf{r}_1' \int d\mathbf{r}_2' G(\mathbf{r}_2, \mathbf{r}_2')$$
$$\times \phi_\alpha^*(\mathbf{r}_1')V_2(\mathbf{r}_1', \mathbf{r}_2')\Psi(\mathbf{r}_1', \mathbf{r}_2') . \tag{3.50}$$

If the explicit form of the Green's function is now used [see (2.13)], then the scattering amplitude has the form

$$f_\alpha(k_0, \theta) = -\frac{1}{2\pi}\int d\mathbf{r}_1' e^{ik_\alpha \mathbf{r}_2'}\phi_\alpha^*(\mathbf{r}_1')V_2(\mathbf{r}_1', \mathbf{r}_2')\Psi(\mathbf{r}_1', \mathbf{r}_2')d\mathbf{r}_2' . \tag{3.51}$$

If one uses the following expression for $\Psi(\mathbf{r}_1', \mathbf{r}_2')$,

$$\Psi(\mathbf{r}_1', \mathbf{r}_2') = \phi_0(\mathbf{r}_1')e^{ik_0 \mathbf{r}_2'} , \tag{3.52}$$

then the Born approximation is obtained.

The exchange in inelastic scattering can be taken into account as was done for elastic processes [compare with (3.17)]:

$$\Psi(\mathbf{r}_1, \mathbf{r}_2) = \frac{1}{(2)^{1/2}}\sum_\alpha \left[F_\alpha^\pm(\mathbf{r}_1)\phi_\alpha(\mathbf{r}_2) \pm F_\alpha^\pm \phi_\alpha(\mathbf{r}_1)\right] , \tag{3.53}$$

where "$\pm$" denotes singlet and triplet scattering, respectively. Substituting (3.53) into the Schrödinger equation we obtain, as usual

$$\left(\nabla_1^2 + k_\alpha^2\right)F_\alpha^\pm(\mathbf{r}_1) = 2\sum_\beta R_{\alpha\beta}^\pm(\mathbf{r}_1) ,$$

where

$$R^{\pm}_{\alpha\beta}(\mathbf{r}_1) = F^{\pm}_{\beta}(\mathbf{r}_1) \int d\mathbf{r}_2 \phi^*_{\alpha}(\mathbf{r}_2) V_1(\mathbf{r}_1,\mathbf{r}_2) \phi_{\beta}(\mathbf{r}_2)$$

$$\pm \phi_{\beta}(\mathbf{r}_1) \int d\mathbf{r}_2 \phi^*_{\alpha}(\mathbf{r}_2) \left[V_2(\mathbf{r}_1,\mathbf{r}_2) - \tfrac{1}{2}\left(\nabla^2_2 + k^2_{\beta}\right) \right] F^{\pm}_{\beta}(\mathbf{r}_2) , \quad (3.54)$$

and $V_1(\mathbf{r}_1,\mathbf{r}_2) = 1/r_{12} - 1/r_1$ is the energy of interaction between electron 1 and the hydrogen atom consisting of electron 2 and the nucleus; $V_2(\mathbf{r}_1,\mathbf{r}_2) = 1/r_{12} - 1/r_2$ is that for electron 2 and the hydrogen atom consisting of electron 1 and the nucleus. The following notation is very frequently used in the literature:

$$V_{\alpha\beta}(\mathbf{r}_1) = \int d\mathbf{r}_2 \phi^*_{\alpha}(\mathbf{r}_2) V_1(\mathbf{r}_1,\mathbf{r}_2) \phi_{\beta}(\mathbf{r}_2)$$

$$= \int \frac{\phi^*_{\alpha}(\mathbf{r}_2)\phi_{\beta}(\mathbf{r}_2)}{|\mathbf{r}_1 - \mathbf{r}_2|} d\mathbf{r}_2 - \frac{\delta_{\alpha\beta}}{r_1} , \quad (3.55)$$

$$W_{\alpha\beta}(\mathbf{r}_1,\mathbf{r}_2) = \phi^*_{\alpha}(\mathbf{r}_2)\phi_{\beta}(\mathbf{r}_1) \left[V_2(\mathbf{r}_1,\mathbf{r}_2) - \frac{\nabla^2_2 + k^2_{\beta}}{2} \right] . \quad (3.56)$$

The scattering amplitude has the form

$$f^{\pm}_{\alpha} = -\frac{1}{2\pi} \sum_{\beta} \int e^{-i\mathbf{k}_{\alpha}\mathbf{r}_1} R^{\pm}_{\alpha\beta}(\mathbf{r}_1) d\mathbf{r}_1 . \quad (3.57)$$

The excitation cross sections σ^+_{α} and σ^-_{α}, which correspond to singlet and triplet scattering, are determined by expressions of the type (3.46), and since they have statistical weights of $\tfrac{1}{4}$ and $\tfrac{3}{4}$, respectively, the differential cross section for the $0-\alpha$ transition has the form:

$$\frac{d\sigma_{\alpha}}{d\Omega} = \frac{1}{4}\frac{d\sigma^+_{\alpha}}{d\Omega} + \frac{3}{4}\frac{d\sigma^-_{\alpha}}{d\Omega} . \quad (3.58)$$

The Born-Oppenheimer approximation is based on the assumption that in the right-hand side of (3.54) we have

$$F^{\pm}_0(\mathbf{r}) = e^{i\mathbf{k}_0\mathbf{r}} ; \quad F^{\pm}_{\alpha}(\mathbf{r}) = 0 ; \quad \alpha \neq 0 . \quad (3.59)$$

One may expect that the application of the distorted-wave approximation that takes account of the partial distortion of both the incident and scattered electron wave will provide higher accuracy. Let us start with the analysis of the two-state approximation again with the ground (0) and final (α) states of a hydrogen atom. Then the infinite system of coupled equations is reduced to, [see (3.54)],

$$\left(\nabla^2_1 + k^2_0\right) F_0(\mathbf{r}_1) = 2(R_{00} + R_{0\alpha}) ;$$
$$\left(\nabla^2_1 + k^2_{\alpha}\right) F_{\alpha}(\mathbf{r}_1) = 2(R_{\alpha 0} + R_{\alpha\alpha}) \quad (3.60)$$

where

$$R_{\alpha\beta}(r_1) = F_\beta(r_1) \int dr_2 \phi_\alpha^*(r_2)(1/r_{12} - 1/r_1)\phi_\beta(r_2)$$

$$\pm \phi_\beta(r_1) \int dr_2 \phi_\alpha^*(r_2) \left(1/r_{12} - 1/r_1 - \frac{\nabla_2^2 + k_\beta^2}{2} \right) F_\beta(r_2) \, .$$

Then the Hermiticity condition of ∇_2^2 may be used.

Remember that $\phi_\alpha(r_2)$ obeys an equation of the type of (3.4). Due to this condition, one has:

$$\left[\nabla_1^2 + k_0^2 - 2V_{00}(r_1) \right] F_0^\pm(r_1) \mp 2 \int W_{00}(r_1, r_2) F_0^\pm(r_2) dr_2$$

$$= 2 \left[V_{0\alpha}(r_1) F_\alpha^\pm(r_1) \pm \int W_{0\alpha}(r_1, r_2) F_\alpha^\pm(r_2) dr_2 \right] \, ;$$

$$\left[\nabla_1^2 + k_\alpha^2 - 2V_{\alpha\alpha}(r_1) \right] F_\alpha^\pm(r_1) \mp 2 \int W_{\alpha\alpha}(r_1, r_2) F_\alpha^\pm(r_2) dr_2$$

$$= 2 \left[V_{\alpha 0}(r_1) F_0^\pm(r_1) \right] \pm \int W_{\alpha 0}(r_1, r_2) F_0^\pm(r_2) dr_2 \, . \tag{3.61}$$

Here $V_{\alpha\beta}(r_1)$ is given by (3.55), and

$$W_{\alpha\beta}(r_1, r_2) = \phi_\alpha^*(r_2)\phi_\beta(r_1)(1/r_{12} + E_\alpha + E_\beta - E) \, . \tag{3.62}$$

This procedure should be carried out with care, however, because here we also meet uncertainties which are similar to those mentioned above. These uncertainties result from the wave functions being known only approximately. Indeed, if the wave function ϕ of the atom is known approximately, $|\tilde\phi - \phi| < g(r)$, where $g(r)$ is a certain function decreasing at infinity, then the transition

$$\int \tilde\phi^2 \nabla^2 F dr \rightarrow \int F \nabla^2 \tilde\phi dr = - \int F \left(-k_0^2 - U \right) \tilde\phi dr$$

is valid within the assumption that the expression of the type $\int F(r)(\Delta + k_0^2 + U)g(r)dr$ may be neglected. The proof of this assumption requires a special study for each case. This problem has been discussed in detail by *Moiseiwitch* [3.7].

Consider the approximate solution to the first equation in (3.61) where the nondiagonal terms in the interaction matrix are omitted,

$$\left[\nabla_1^2 + k_0^2 - 2V_{00}(r_1) \right] F_0^\pm(r_1) \mp 2 \int W_{00}(r_1, r_2) F_0^\pm(r_2) dr_2 = 0 \, .$$

The solution of this equation was discussed in detail in the preceding section. Note that this solution can be found only numerically. Substituting $F_0(r)$ into the second equation of (3.61) and requiring the asymptote of $F^\pm$ to be $F^\pm(r \rightarrow \infty) \sim f_\alpha^\pm(k_0, \theta)r^{-1} \exp(ik_\alpha r)$, we obtain the wave function $F_\alpha^\pm(r)$. Then the obtained $F_0^\pm(r)$ and $F_\alpha^\pm(r)$ must be substituted into the scattering amplitude:

$$f_\alpha^\pm = -\frac{1}{2\pi} \int d\mathbf{r}_1 F_\alpha^\pm(\mathbf{r}_1) \Big[V_{\alpha 0}(\mathbf{r}_1) F_0^\pm(\mathbf{r}_1)$$

$$\pm \int d\mathbf{r}_2 W_{\alpha 0}(\mathbf{r}_1, \mathbf{r}_2) F_0^\pm(\mathbf{r}_2) \Big] . \tag{3.63}$$

This method of obtaining the amplitude is one of the realizations of the distorted-wave method.

There exist a number of various modifications of this method [3.8]. For example, a partial-wave expansion of $F_0^\pm(\mathbf{r})$ and $F_\alpha^\pm(\mathbf{r})$ can be made. In this case the total excitation cross section σ_α is expressed in terms of partial cross sections. The distorted-wave method is quite suitable for the description of atomic excitation. Even more precise, however, is the close-coupling method to be discussed in the following section.

3.4 Close-Coupling Method

System (3.61) is an example of the use of the close-coupling method (CCM) for the case when only two channels are included, 0 and α. If the series (3.53) is not truncated, then we obtain an infinite system of equations that can not be solved. Therefore, in practice, only low-lying states are included in the wave function expansion which results in a finite system of coupled equations.

If an arbitrary number of channels is involved, the system of equations is written, similarly to (3.61) as,

$$\left(\nabla_1^2 + k_\alpha^2 \right) F_\alpha^\pm(\mathbf{r}_1)$$

$$= 2 \sum_\beta \left[V_{\alpha\beta}(\mathbf{r}_1) F_\beta^\pm(\mathbf{r}_1) \pm \int d\mathbf{r}_2 W_{\alpha\beta}(\mathbf{r}_1, \mathbf{r}_2) F_\beta^\pm(\mathbf{r}_2) \right] . \tag{3.64}$$

To obtain the radial equations, we shall expand $F_\alpha^\pm(\mathbf{r}_1)$ into a series over the partial functions. To do that, one may use the expansion (3.14), but owing to the fact that the real radial functions are more convenient, we prefer to utilize the following expansion:

$$F_\alpha^\pm(\mathbf{r}_1) = \sum_{lm} \frac{f_{\alpha lm}^\pm(r_1)}{r_1} Y_{lm}(\Omega_1) . \tag{3.65}$$

Substitution of (3.65) into (3.64) gives:

$$\sum_{lm} \left[\frac{1}{r_1} \left(\frac{d^2}{dr_1^2} - \frac{l(l+1)}{r_1^2} + k_\alpha^2 \right) f_{\alpha lm}^\pm(r_1) Y_{lm}(\Omega_1) \right]$$

$$= \sum_{\beta l'm'} 2 \left[\frac{f_{\beta l'm'}^\pm(r_1)}{r_1} V_{\alpha\beta}(\mathbf{r}_1) Y_{l'm'}(\Omega_1) \right.$$

$$\left. \pm \int d\mathbf{r}_2 W_{\alpha\beta}(\mathbf{r}_1, \mathbf{r}_2) \frac{f_{\beta l'm'}^\pm(r_2)}{r_2} Y_{l'm'}(\Omega_2) \right] . \tag{3.66}$$

33

In the right-hand side of (3.66) we have taken into account that the principal quantum number n may not coincide with n' in the left-hand side, which leads to the angular orbital momentum l' being in general different from l. We shall below use the following notation:

$$\frac{d^2}{dr^2} - \frac{l(l+1)}{r_2} + k_\alpha^2 \equiv D_{\alpha l}(r) .$$
(3.67)

We multiply (3.66) by $Y^*_{l''m''}(\Omega_1)$ and integrate it over Ω_1 using the condition of orthogonality for the spherical functions. We obtain the system of CCM equations for radial functions:

$$D_{\alpha l''}(r_1)f^\pm_{\alpha l''m''}(r_1) = \sum_{\beta l'm'} \left[V^{l''m''l'm'}_{\alpha\beta}(r_1)f^\pm_{\beta l'm'}(r_1) \right.$$

$$\left. \pm \int r_1 r_2 dr_2 W^{l''m''l'm'}_{\alpha\beta}(r_1, r_2)f^\pm_{\beta l'm'}(r_2) \right],$$

where

$$V^{l''m''l'm'}_{\alpha\beta}(r_1) = 2 \int d\Omega_1 Y^*_{l''m''}(\Omega_1) V_{\alpha\beta}(r_1) Y_{l'm'}(\Omega_1) ;$$

$$W^{l''m''l'm'}_{\alpha\beta}(r_1, r_2) = 2 \iint d\Omega_1 d\Omega_2 Y^*_{l''m''}(\Omega_1) W_{\alpha\beta}(r_1, r_2) Y_{l'm'}(\Omega_2) .$$
(3.68)

As can be seen, the obtained expressions are quite complicated. The system (3.68) is, unfortunately, even more combersome than it looks at first sight: in the right-hand side of (3.68) the unknown function $f^\pm(r)$ appears with subscripts $\beta l'm'$ which run over all possible values. To avoid this, the total angular momentum representation is used, since this allows one to remove the unlimited summation over l' in the right-hand side of the corresponding system.

3.5 Total Angular Momentum Representation

Let us return to the expansion of the total wave function $\Psi(r_1, r_2)$ [see (3.44)]. Since the subscript α describes a certain state, it contains a combination of three quantum numbers $(n \; l_1 \; m_1)$ which describe the state of an orbital electron. Combining (3.44) with (3.65),

$$\Psi(r_1, r_2) = \sum_{nl_1m_1l_2m_2} \frac{f^\pm_{nl_1m_1l_2m_2}(r_1)}{r_1} Y_{l_2m_2}(\Omega_2)\phi_{nl_1m_1}(r_2) .$$
(3.69)

This representation, as stated above, is inconvenient to use. Proceeding from the expansion (3.69) we can introduce another radial function characterized by the L and M values, which is achieved by using the Clebsch-Gordan coefficients (with the "$\pm$" signs omitted provisionally)

$$f^{LM}_{nl_1l_2}(r) = \sum_{m_1m_2} C^{LM}_{l_1m_1l_2m_2} f_{nl_1m_1l_2m_2}(r) ,$$
(3.70)

where

$$L = l_1 + l_2 ; \quad M = m_1 + m_2 .$$

Now we examine a few relations which will be used below. We shall need to invert (3.70) making use of the orthogonality of the Clebsch-Gordan coefficients:

$$\sum_{LM} C^{LM}_{l_1 m_1 l_2 m_2} C^{LM}_{l_1 m_1' l_2 m_2'} = \delta_{m_1 m_1'} \delta_{m_2 m_2'} . \tag{3.71a}$$

We also shall employ hereafter the following orthogonality condition:

$$\sum_{m_1 m_2} C^{LM}_{l_1 m_1 l_2 m_2} C^{L'M'}_{l_1 m_1 l_2 m_2} = \delta_{LL'} \delta_{MM'} . \tag{3.71b}$$

Multiplication of (3.70) by $C^{LM}_{l_1 m_1' l_2 m_2'}$ and further summation over LM lead, using (3.71a), to

$$f_{nl_1 m_1 l_2 m_2}(r) = \sum_{LM} C^{LM}_{l_1 m_1 l_2 m_2} f^{LM}_{nl_1 l_2}(r) . \tag{3.72}$$

The expressions for the potentials V and W are obtained similarly. In the total angular momentum representation, we have:

$$V^{LML'M'}_{l_1 l_1' l_2 l_2'}(r) = \sum_{m_1 m_2 m_1' m_2'} C^{LM}_{l_1 m_1 l_2 m_2} V^{l_2 m_2 l_2' m_2'}_{nl_1 m_1 l_2 m_2}(r) C^{L'M'}_{l_1' m_1' l_2' m_2'} . \tag{3.73}$$

In analogy with (3.72) we obtain the inverse relation:

$$V^{l_2 m_2 l_2' m_2'}_{nl_1 m_1 n' l_1' m_1'}(r) = \sum_{LML'M'} C^{LM}_{l_1 m_1 l_2 m_2} V^{LML'M'}_{l_1 l_1' l_2 l_2'}(r) C^{L'M'}_{l_1' m_1' l_2' m_2'} . \tag{3.74}$$

Here an important advantage of the total angular momentum representation is utilized, namely, because of conservation of total momentum, only those transitions for which $L = L'$ and $M = M'$ occur. Then (3.74) is simplified to:

$$V^{l_2 m_2 l_2' m_2'}_{nl_1 m_1 n' l_1' m_1'}(r) = \sum_{LM} C^{LM}_{l_1 m_1 l_2 m_2} V^{LM}_{l_1 l_1' l_2 l_2'}(r) C^{LM}_{l_1' m_1' l_2' m_2'} . \tag{3.75}$$

Note that with the incident and atomic-electron subscripts included explicitly, (3.68) becomes

$$D_{\alpha l_2}(r) f_{\underbrace{nl_1 m_1}_{\alpha} l_2 m_2}(r)$$

$$= \sum_{\substack{n' l_1' m_1' \\ l_2' m_2'}} V^{l_2 m_2 l_2' m_2'}_{nl_1 m_1 n' l_1' m_1'}(r) f_{n' l_1' m_1' l_2' m_2'}(r) + \text{the similar term with } W .$$

$$\tag{3.76}$$

We multiply both sides of (3.76) by $C_{l_1 m_1 l_2 m_2}^{LM}$ and sum over m_1, m_2. Then

$$\sum_{m_1 m_2} C_{l_1 m_1 l_2 m_2}^{LM} D_{\alpha l_2}(r) f_{n l_1 m_1 l_2 m_2}(r) = \sum_{n' l_1' m_1' l_2' m_2'} \sum_{m_1 m_2} \sum_{L' M'}$$

$$\times C_{l_1 m_1 l_2 m_2}^{LM} C_{l_1 m_1 l_2 m_2}^{L' M'} C_{l_1' m_1' l_2' m_2'}^{L' M'} V_{l_1 l_1' l_2 l_2'}^{L' M'}(r) f_{n' l_1' m_1' l_2' m_2'} + \cdots \qquad (3.77)$$

where (3.75) has been used. According to (3.70), the left-hand side contains the radial wave function in the total angular momentum representation, thus:

$$D_{\alpha l_2}(r) f_{n l_1 l_2}^{LM}(r) = \sum_{\substack{n' l_1' m_1' l_2' m_2' \\ m_1 m_2 L' M'}} C_{l_1' m_1' l_2' m_2'}^{L' M'} C_{l_1 m_1 l_2 m_2}^{LM} C_{l_1 m_1 l_2 m_2}^{L' M'}$$

$$\times V_{l_1 l_1' l_2 l_2'}^{L' M'}(r) f_{n' l_1' m_1' l_2' m_2'}(r) + \cdots \qquad (3.78)$$

The orthogonality condition (3.71a) allows one to sum over $L' M'$

$$D_{\alpha l_2}(r) f_{n l_1 l_2}(r) = \sum_{n' l_1' m_1' l_2' m_2'} C_{l_1' m_1' l_2' m_2'}^{LM} V_{l_1 l_1' l_2 l_2'}^{LM}(r) f_{n' l_1' m_1' l_2' m_2'}(r) + \cdots . \qquad (3.79)$$

Using now (3.70) we obtain finally:

$$D_{\alpha l_2}(r_1) f_{n l_1 l_2}^{\pm LM}(r_1) = \sum_{n' l_1' l_2'} V_{l_1 l_1' l_2 l_2'}^{LM}(r_1) f_{n' l_1' l_2'}^{\pm LM}(r_1)$$

$$\pm \sum_{n' l_1' l_2'} \int r_1 r_2 dr_2 W_{l_1 l_1' l_2 l_2'}^{LM}(r_1, r_2) f_{n' l_1' l_2'}^{\pm LM}(r_2) . \qquad (3.80)$$

It is obvious that the system of equations (3.80) in the total angular momentum representation does not couple radial functions with different L and M.

3.6 Two-State Approximation Within the Close-Coupling Method

The system of equations (3.80) can be specified and simplified with the use of the explicit form for the potentials V and W. Let us rewrite the potential in the total angular momentum representation, taking into account the orthogonality condition. Then we obtain instead of (3.73)

$$V_{l_1 l_1' l_2 l_2'}^{LM}(r) = \sum_{m_1 m_2 m_1' m_2'} C_{l_1 m_1 l_2 m_2}^{LM} C_{l_1' m_1' l_2' m_2'}^{LM} V_{n l_1 m_1 n' l_1' m_1'}^{l_2 m_2 l_2' m_2'}(r) . \qquad (3.81)$$

The wave functions for the hydrogen atoms are represented, as usual, in the form

$$\phi_{n l_1 m_1}(r_2) = R_{n l_1}(r_2) Y_{l_1 m_1}(\Omega_2) . \qquad (3.82)$$

Substituting (3.82) into (3.54) and using (3.9) we get

$$V_{\alpha\beta}(\mathbf{r}_1) = 2 \sum_{\lambda} \int r_2^2 dr_2 \gamma_\lambda R_{nl_1}(r_2) R_{n'l_1'}(r_2)$$

$$\times \int d\Omega_2 Y_{l_1 m_1}^*(\Omega_2) P_\lambda(c) Y_{l_1' m_1'}(\Omega_2) - 2\delta_{\alpha\beta}/r_1 , \qquad (3.83)$$

Where $\gamma_\lambda = r_<^\lambda / r_>^{\lambda+1}$. To make this formula more compact, we denote the radial part as:

$$(\gamma_\lambda)_{nl_1 n'l_1'} = \int r_2^2 dr_2 \gamma_\lambda R_{nl_1}(r_2) R_{n'l_1'}(r_2) . \qquad (3.84)$$

Then, according to (3.68), the potential becomes

$$V_{l_1 l_1' l_2 l_2'}^{LM}(r) = \sum_{m_1 m_2 m_1' m_2'} C_{l_1 m_1 l_2 m_2}^{LM} C_{l_1' m_1' l_2' m_2'}^{LM}$$

$$\times 2 \int d\Omega_1 Y_{l_2 m_2}^*(\Omega_1) Y_{l_2' m_2'}(\Omega_1) \left[\sum_\lambda (\gamma_\lambda)_{nl_1 n'l_1'} \right.$$

$$\times \left. \int d\Omega_2 Y_{l_1 m_1}^*(\Omega_2) P_\lambda(c) Y_{l_1' m_1'}(\Omega_2) - \delta_{\alpha\beta}/r \right] . \qquad (3.85)$$

The last term $\delta_{\alpha\beta}/r$ is, in fact, equal to $\delta_{nn'}\delta_{l_1 l_1'}\delta_{m_1 m_1'}/r$. The orthogonality condition for spherical functions gives $\delta_{l_2 l_2'}\delta_{m_2 m_2'}$. Thus, the last term equals $-1/r$. Then (3.85) may be rewritten in reduced form:

$$V_{l_1 l_1' l_2 l_2'}^{LM}(r) = \sum_\lambda (\gamma_\lambda)_{nl_1 n'l_1'} h_{\lambda l_1 l_1' l_2 l_2'}^{LM} - \frac{2}{r} , \qquad (3.86)$$

$$h_{\lambda l_1 l_1' l_2 l_2'}^{LM} = \sum_{m_1 m_2 m_1' m_2'} C_{l_1 m_1 l_2 m_2}^{LM} C_{l_1' m_1' l_2' m_2'}^{LM}$$

$$\times 2 \int d\Omega_1 Y_{l_2 m_2}^*(\Omega_1) Y_{l_2' m_2'}(\Omega_1) \int d\Omega_2 Y_{l_1 m_1}^*(\Omega_2)$$

$$\times P_\lambda(c) Y_{l_1' m_1'}(\Omega_2) . \qquad (3.87)$$

As has been shown in [3.9], h^{LM} may be expressed through Racah coefficients

$$h_{\lambda l_1 l_1' l_2 l_2'}^{LM} = (-1)^{l_1 + l_1' - L}(2\lambda + 1)^{-1} [(2l_1 + 1)$$

$$\times (2l_1' + 1)(2l_2 + 1)(2l_2' + 1)]^{1/2} C_{l_1 0 l_1' 0}^{\lambda 0} C_{l_2 0 l_2' 0}^{\lambda 0}$$

$$\times W(\lambda l_1 l_1' l_2 l_2' L) , \qquad (3.88)$$

where $W(abcdef)$ are the Racah coefficients.

Consider now the term with W. Combination of (3.62) with (3.82) gives

$$W_{\alpha\beta}(\mathbf{r}_1, \mathbf{r}_2) = R_{nl_1}(r_2) Y_{l_1 m_1}^*(\Omega_2) R_{n'l_1'}(r_1)$$

$$\times Y_{l_1' m_1'}(\Omega_1)(1/r_{12} + E_n + E_{n'} - E) ;$$

$$W_{nl_1 m_1 n'l_1' m_1'}^{l_2 m_2 l_2' m_2'}(r_1, r_2) = 2 \int d\Omega_1 \int d\Omega_2 R_{nl_1}(r_2)$$

$$\times R_{n'l_1'}(r_1)Y_{l_1 m_1}^*(\Omega_2)Y_{l_2' m_2'}(\Omega_2)Y_{l_1' m_1'}(\Omega_1) \qquad (3.89)$$

$$\times Y_{l_2 m_2}^*(\Omega_1)\left[\sum_\lambda \gamma_\lambda P_\lambda(c) + E_n + E_{n'} - E)\right].$$

In the total angular momentum representation, $W_{l_1 l_1' l_2 l_2'}^{LM}$ may be expressed similarly to (3.81)

Let us introduce the following notation:

$$r_1 r_2 R_{nl_1}(r_2)R_{n'l_1'}(r_1) \equiv (\gamma_0^l)_{nl_1 n'l_1'};$$

$$\gamma_\lambda r_1 r_2 R_{nl_1}(r_2)R_{n'l_1'}(r_1) \equiv (\gamma_\lambda^l)_{nl_1 n'l_1'};$$

$$\sum_{m_1 m_2 m_1' m_2'} C_{l_1 m_1 l_2 m_2}^{LM} C_{l_1' m_1' l_2' m_2'}^{LM} \int d\Omega_1 \int d\Omega_2 Y_{l_2 m_2}^*(\Omega_1)$$

$$\times P_\lambda(c)Y_{l_2' m_2'}(\Omega_2)Y_{l_1 m_1}^*(\Omega_2)Y_{l_1' m_1'}(\Omega_1) \equiv g_{\lambda l_1 l_1' l_2 l_2'}^{LM}. \qquad (3.90)$$

Similarly to (3.88), g^{LM} can be calculated explicitly

$$g_{\lambda l_1 l_1' l_2 l_2'}^{LM} = (-1)^{l_2+l_2'}(2\lambda+1)^{-1}\left[(2l_1+1)(2l_1'+1)(2l_2+1)(2l_2'+1)\right]^{1/2}$$

$$\times C_{l_1 0 l_2' 0}^{\lambda 0}C_{l_2 0 l_1' 0}^{\lambda 0}W(\lambda l_1 l_2 l_2' l_1' L). \qquad (3.91)$$

Upon substitution of these expressions into (3.80) we obtain the system of integro-differential equations for the radial functions in the total angular momentum representation (with the spin subscripts "$\pm$" restored)

$$\left(\frac{d_2}{dr_1^2} - \frac{l_2(l_2+1)}{r_1^2} + k_n^2 + 2/r_1\right)f_{nl_1 l_2}^{\pm LM}(r_1)$$

$$= \sum_{n'l_1' l_2'\lambda} f_{n'l_1' l_2'}^{\pm LM}(r_1)h_{\lambda l_1 l_1' l_2 l_2'}^{LM}(\gamma_\lambda)_{nl_1 n'l_1'}$$

$$\pm \int dr_2 f_{n'l_1' l_2'}^{\pm LM}(r_2)2\left[(\gamma_\lambda')_{nl_1 n'l_1'}g_{\lambda l_1 l_1' l_2 l_2'}^{LM}\right.$$

$$\left. + (\gamma_0')_{nl_1 n'l_1'}g_{0l_1 l_1' l_2 l_2'}^{LM}(E_n + E_{n'} - E)\right]. \qquad (3.92)$$

Although this system looks rather complicated, it is in fact quite simple. The point is that the momenta obey the so-called triangle rules, which simplifies the system significantly. Therefore, $h_\lambda \neq 0$ only if

$$|l_1 - l_2| \leq L \leq l_1 + l_2; \qquad |l_1' - l_2'| \leq L \leq l_1' + l_2';$$
$$|l_1' - l_1| \leq \lambda \leq l_1' + l_1; \qquad |l_2' - l_2| \leq \lambda \leq l_2' + l_2;$$
$$(-1)^{l_1+l_2} = (-1)^{l_1'+l_2'},$$

where $l_1 + l_2 + \lambda$ is even. Similar rules are valid for g_λ.

Consider, for example, the system of equations for the 1s − 2s transition (e + H → e + H* processes). Since in our notation the subscript "1" denotes the atomic electron, the chosen transition means that $l_1 = l'_1 = 0$. Then it follows from the above relations that $\lambda = 0$, $l_2 = l'_2 = L$. Hence (3.92) is reduced to a system of two integro-differential equations:

$$\left(\frac{d^2}{dr^2} - \frac{L(L+1)}{r^2} + k_1^2 + \frac{2}{r}\right) f_{10L}^{\pm L}(r)$$

$$= f_{10L}^{\pm L}(r)(\gamma_0)_{1010} h_{000LL}^L + f_{20L}^{\pm L}(r)(\gamma_0)_{1020} h_{000LL}^L$$

$$\pm 2 \int dr' \left\{ f_{10L}^{\pm L}(r') g_{L00LL}^L \left[(\gamma'_L)_{1010} + (\gamma_0)_{1010}(2E_1 - E)\delta_{L0} \right] \right.$$

$$\left. + f_{20L}^{\pm L}(r') g_{L00LL}^L \left[(\gamma'_L)_{1020} + (\gamma'_0)_{1020}(E_1 + E_2 - E)\delta_{L0} \right] \right\} \ ,$$

$$\left[\frac{d^2}{dr^2} - \frac{L(L+1)}{r^2} + k_2^2 + \frac{2}{r}\right] f_{20L}^{\pm L}(r)$$

$$= f_{10L}^{\pm L}(r)(\gamma_0)_{2010} h_{000LL}^L + f_{20L}^{\pm L}(r)(\gamma_0)_{2020} h_{000LL}^L$$

$$\pm \int dr' 2 \left\{ f_{10L}^{\pm L}(r') g_{L00LL}^L \left[(\gamma'_L)_{2010} + (\gamma'_0)_{2010}(E_1 + E_2 - E)\delta_{L0} \right] \right.$$

$$\left. + f_{20L}^{\pm L}(r') g_{L00LL}^L \left[(\gamma'_L)_{2020} + (\gamma'_0)_{2020}(2E_2 - E)\delta_{L0} \right] \right\} \ . \tag{3.93}$$

Let us omit the subscript M since the equations are independent of it. The coefficients h_λ and g_λ can be easily obtained by (3.88,91).

The number of partial waves involved depends on the convergence of the series (3.69). In the considered case of low-energy electron–hydrogen-atom scattering (the collision energy being less than, e.g., 1 Ry), one may confine oneself to $L \le 3 - 4$. For s - s transitions the total angular momentum representation provides no advantages: in both cases the system contains only two amplitudes, f_{10L} and f_{20L}. For the transitions with nonzero l_1 (or l'_1), however, the triangle rules will truncate the sum in the right-hand side of (3.92) only within the total angular momentum representation. The potentials for hydrogen can be calculated in the explicit analytic form. From (3.84) one may obtain:

$$(\gamma_0)_{1010} = -e^{-2r}(1 + 1/r) + 1r \ ,$$

$$(\gamma_0)_{1020} = (\gamma_0)_{2010} = \frac{2\sqrt{2}}{9} e^{-3r/2}(r + 2/3) \ , \tag{3.94}$$

$$(\gamma_0)_{2020} = e^{-r}(r^2/8 + r/4 + 3/4 + 1/r) + 1/r \ .$$

Let us say a few words about the properties of the solutions of the system (3.93). It is known from the theory of differential equations that a system of n second-order equations has $2n$ linearly indepent solutions. Thus in our case the system (3.93) has four linearly independent solutions. As $r \to 0$, the system (3.93) is split into two equations of the form (3.16) with $U = -2Z/r$, i.e., into the equations for an electron in a coulomb field, each of them having two linearly independent solutions

$$f_{1L}^{(1)}(r \to 0) \sim r^{L+1} \ , \quad f_{1L}^{(3)}(r \to 0) \sim r^{-L} \ ,$$

$$f_{2L}^{(2)}(r \to 0) \sim r^{L+1} \ , \quad f_{2L}^{(4)}(r \to 0) \sim r^{-L} \ .$$

Here we have introduced the simplified notation ($f_{10L}^{\pm L} = f_{1L}$, etc.) The general solution of the system (3.93) is composed of these partial solutions (at $r \to 0$).

$$f_L \equiv \begin{pmatrix} f_{1L} \\ f_{2L} \end{pmatrix} = C_1 \begin{pmatrix} f_{1L}^{(1)} \\ 0 \end{pmatrix} + C_2 \begin{pmatrix} 0 \\ f_{2L}^{(1)} \end{pmatrix} + C_3 \begin{pmatrix} f_{1L}^{(3)} \\ 0 \end{pmatrix} + C_4 \begin{pmatrix} 0 \\ f_{2L}^{(4)} \end{pmatrix} . \tag{3.95}$$

From physical considerations requiring the solution to be finite at zero, we have $f_L(r = 0) = \begin{pmatrix} 0 \\ 0 \end{pmatrix}$, and, hence; $C_3 = C_4 = 0$. The remaining two coefficients are determined from the asymptotic conditions. The latter, presented in the form given in (2.33a) (i.e., in the K-matrix form, according to Chap. 2) are more convenient for numerical calculations. Here we have two different possibilities:

i) the electron is falling onto the target in the ground state

$$\begin{pmatrix} \sin(k_1 r - \pi L/2) & + & K_{11} \cos(k_1 r - \pi L/2) \\ & & K_{12} \cos(k_2 r - \pi L/2) \end{pmatrix} , \tag{3.96a}$$

ii) the electron is falling onto the target in the excited state:

$$\begin{pmatrix} & & K_{21} \cos(k_1 r - \pi L/2) \\ \sin(k_2 r - \pi L/2) & + & K_{22} \cos(k_2 r - \pi L/2) \end{pmatrix} , \tag{3.96b}$$

Here the following notation is introduced: the first subscript specifies the atom in the initial state; while the second describes it in the final state. The first possibility is usually that observed in experiments. However, one should know the total K-matrix when carrying out the numerical calculations,

$$K = \begin{pmatrix} K_{11} & K_{12} \\ K_{21} & K_{22} \end{pmatrix}$$

For this reason the calculations are performed with both asymptotics. This also is essential since the K-matrix must be symmetric according to physical considerations. Hence, the requirement of K_{21} being equal to K_{12} allows us to control the numerical results obtained.

In this case, as in the single-channel case, the number of unknowns is greater than the number of equations. Indeed, equating (3.95) and (3.96a), i.e.,

$$C_1 \begin{pmatrix} f_{1L}^{(1)} \\ f_{2L}^{(1)} \end{pmatrix} + C_2 \begin{pmatrix} f_{1L}^{(2)} \\ f_{2L}^{(2)} \end{pmatrix}$$

$$= \begin{pmatrix} \sin(k_1 r - \pi L/2) & + & K_{11} \cos\left(k_1 r - \dfrac{\pi L}{2}\right) \\ & & K_{12} \cos\left(k_2 r - \dfrac{\pi L}{2}\right) \end{pmatrix} , \tag{3.97}$$

we obtain only two equations for four unknowns, C_1, C_2, K_{11} and K_{12}. Thus, one must again use additional physical considerations: the requirements of smooth matching of f_L-functions with asymptotic solutions, i.e., the requirement that their derivatives are identical:

$$C_1 \begin{pmatrix} f_{1L}^{(1)} \\ f_{2L}^{(1)} \end{pmatrix}' + C_2 \begin{pmatrix} f_{1L}^{(2)} \\ f_{2L}^{(2)} \end{pmatrix}'$$

$$= \begin{pmatrix} k_1 \cos(k_1 r - \pi L/2) & -K_{11} k_1 \sin\left(k_1 r - \dfrac{\pi L}{2}\right) \\ & -K_{12} k_2 \cos\left(k_2 r - \dfrac{\pi L}{2}\right) \end{pmatrix}. \tag{3.98}$$

The system of four equations (3.97, 98) allows one to determine C_1, C_2, K_{11}, and K_{12}. A similar system with the second asymptotic condition allows one to obtain the general solution of this problem.

Now consider physical observables. Without going into detail, we point out that the K-matrix is related to the S-matrix by the following expression:

$$S = (1 + iK)(1 - iK)^{-1}, \tag{3.99a}$$

whereas the S-matrix, in turn, is directly related to the cross section:

$$\sigma_{ij} = \frac{1}{k_1^2} \left| S_{ij} - \delta_{ij} \right|^2. \tag{3.99b}$$

We note the following: in the two-channel case the S_{11}-matrix element is different from that obtained for elastic scattering in the single-channel case below the inelastic threshold.

One more remark: from the invariance of the interaction with respect to time reversal it follows that the S-matrix is symmetric [3.4]. Consequently, in the two-channel case only three of the four matrix elements are independent. These three parameters are usually chosen as:

$$S = \begin{pmatrix} \cos \varepsilon \; e^{2i\delta_1} & i \sin \varepsilon \; e^{i(\delta_1 + \delta_2)} \\ i \sin \varepsilon \; e^{i(\delta_1 + \delta_2)} & \cos \varepsilon \; e^{2i\delta_2} \end{pmatrix}.$$

Here δ_1 and δ_2 are the so-called scattering eigenphases for the first and second channels, respectively; ε is a mixing parameter.

It is known that the scattering eigenphases δ_i obtained in the CCM are lower bounds for their exact values provided that the expansion in (3.44) includes all the channels open at a given energy [3.10]. Therefore, the actual calculations should be performed with at least three states taken into account simultaneously (1s, 2s, 2p). The corresponding computations are, however, quite complicated, so we shall omit them here.

Let us consider the results of the calculations of elastic electron–hydrogen-atom scattering cross sections. Figure 3.1 presents the data on the 2s-metastable state excitation. The calculations by CCM were performed with account of the 1s-2s and 1s-2p channels (curve 1) [3.10, 11]. The data obtained are quite surprising: the best results in the entire energy range are given by the Born approximation. This result is accidental, however, whereas more exact calculations by CCM, which take into account six channels, are in excellent agreement with ex-

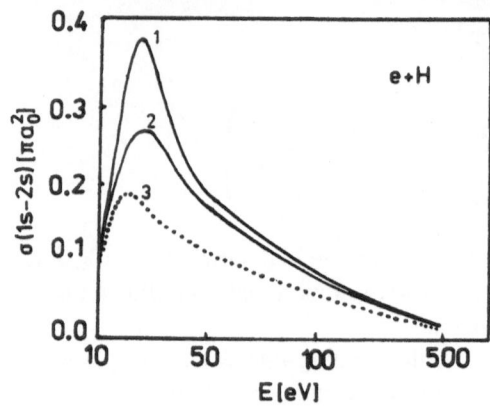

Fig. 3.1. Total cross section for the electron-impact excitation of the 2s level of the H atom. *1* – CCM calculation, *2* – Born approximation, *3* – experiment

periments. Moreover, the inclusion of closed channels results in the appearance of resonances. The cross section in the resonance region may be several times, even by an order of magnitude, larger than the cross section for potential scattering. Thus, resonance effects often play a major role in the scattering. Chapters 4–6 deal with this in detail.

4. Single-Channel Theory of Resonance Scattering

The present chapter deals with the interaction of the autoionizing state (AIS) with the continuum which results in the appearance of resonances in the scattering. The methods of calculation of the AIS parameters which are important for the analysis of the resonance structure in the scattering cross sections are explained. Besides the widely known methods currently used, the original modification of Feshbach's method – the diagonalization method (DM) – is discussed. The DM is used to obtain quantitative results on the resonances in the elastic cross section for electron scattering by an He^+ ion.

All the material considered here is related to the two-channel approximation in which the single open (elastic) and the single closed (inelastic) channels are taken into account.

4.1 Closed-Channel Resonances. Fano's Theory

As has been shown in Sect. 2.5, scattering in a potential field can give rise to shape resonances. It appears that the same resonances may be observed also in the electron scattering by atoms. In addition to these, other types of the resonances which arise in electron scattering by atoms (and, similarly, in scattering of nuclear particles by nuclei, photoexcitation, ionization, etc.) result from the formation of short-lived states of the system "incident particle + target". Since this state decays after a short period of time, it is said that resonances of this type result from the interaction of discrete states with the continuum, and exactly these discrete short-lived states are AIS. In other words, an incident particle excites a compound target and is captured by the latter for a certain time. It is just this time delay which appears as a resonance in the cross section. Resonances of this type have been studied thoroughly by *Feshbach, Goldberger, Fano* [4.1–3], and others.

Consider the definition of AIS in detail. Suppose that the total Hamiltonian can be represented in the form $H = H_0 + V$, where H_0 is a certain unperturbed Hamiltonian, V is a perturbation. The situation may arise when H_0, besides the pure discrete and continuous spectra, has a discrete one submerged in the continuous spectrum. As an example, we can take the case of doubly excited $2sn'l'$-states ($n' \geq 2$) of an He atom, the energies of which coincide with those of $1s\varepsilon l'$-states in the continuous spectrum of the system (e + He^+). In the $1s\varepsilon l'$-state, the coupled electron is in the 1s-state; while the other is in the continuous spectrum having momentum l'.

The discrete states with energies in the continuum are eigenvalues of a certain unperturbed Hamiltonian H_0, for which the Schrödinger equation may be solved relatively easily. These states Ψ_n of the discrete spectrum of the H_0-operator whose energies lie in the continuous spectrum of the unperturbed Hamiltonian are called AIS. The same states exist for the negative ions as well, and they are usually called the autodetaching states. For the sake of simplicity we shall only use the AIS notation, assuming that the experienced reader will not be misled by this term.

It should be noted that AIS are not eigenstates of the total Hamiltonian $H = H_0 + V$. Under the influence of the perturbation V, AIS must decay during some characteristic time [see (2.47)]. Splitting of H into H_0 and V is ambiguous: for example, H_0 may be taken to be the Hamiltonian in the noninteracting electron approximation as well as the Hartree-Fock Hamiltonian. Thus, generally speaking, certain ambiguities may arise in determining the parameters of AIS. We may eliminate these ambiguities if we pass from the independent-electron model to the multiconfiguration wave functions Φ_n, which are constructed as linear combinations

$$\Phi_n = \sum_m \beta_{nm} \Psi_m , \tag{4.1}$$

where the configuration-mixing coefficients β_{nm} are found from the condition that the Φ_n-functions must diagonalize the Hamiltonian H

$$\langle \Phi_n | H | \Phi_{n'} \rangle = \mathcal{E}_n \delta_{nn'} . \tag{4.2}$$

The system of equations (4.1, 2) forms the basis of the so-called configuration-interaction (CI) method of AIS calculation.

U. Fano has shown that the discrete level present in the continuum gives rise to an additional phase shift in the scattering. Consider the most simple case when a single continuum and a single discrete level are present. A generalization of this problem to the multilevel case is not difficult.

Let $|\Phi_0\rangle$ be the wave function of AIS which describes a single discrete level, and let $|\psi(E)\rangle$ describe the state with energy E in the continuous spectrum. We assume these function to be normalized and orthogonal: $\langle \Phi_0 | \Phi_0 \rangle = 1$; $\langle \Phi_0 | \psi(E) \rangle = 0$; $\langle \psi(E) | \psi(E') \rangle = \delta(E - E')$. Let

$$\langle \Phi_0 | H | \Phi_0 \rangle = E_0 , \tag{4.3}$$

where H is the Hamiltonian of the system.

Next, we introduce the following notation:

$$\langle \psi(E) | H | \Phi_0 \rangle = V(E) , \tag{4.4}$$

and impose the following condition:

$$\langle \psi(E') | H | \psi(E) \rangle = E \delta(E - E') . \tag{4.5}$$

Determination of the wave functions $|\Phi_0\rangle$ and $|\psi\rangle$ obeying the above conditions is called the diagonalization of H on the subspaces of discrete and continuous spectra.

Let us now diagonalize H in the space defined by $|\Phi_0\rangle$ and $|\psi\rangle$ by introducing the following eigenfunction:

$$|\Psi(E)\rangle = a(E)|\Phi_0\rangle + \int dE' b(E')|\psi(E')\rangle \ . \tag{4.6}$$

This expansion resembles the one in the CCM when only one term is left in both the discrete and continuous spectra. In this case the wave functions $|\Phi_0\rangle$ and $|\psi(E)\rangle$ describe in general the system "electron + atom" and depend both on the atomic-electron coordinates r_1 and the coordinates of an incident electron, r_2. Hence the expansion coefficients $a(E)$ and $b(E)$ do not depend on coordinates. In the simplest scattering case, integro-algebraic equations are obtained for these coefficients. The diagonalization of H on the subspace of $|\psi(E)\rangle$-states is the most difficult part of the task, though the determination of $|\Phi_0\rangle$ is not a trivial problem too. We shall below give (Sects. 4.3–5) the ways for solving this problem. However, if one considers this task as being solved, then, according to Fano, resonances due to configuration interaction should appear.

Thus, we shall try to solve the Schrödinger equation: $H\Psi = E\Psi$ (3.1) under certain assumptions about the form of Ψ. Let Ψ^N be an approximate solution to this equation, and we shall look for it in the form

$$\Psi^N = \sum_{\alpha}^{N} C_{\alpha}^{N} \psi_{\alpha} \ , \tag{4.7}$$

where ψ_{α} are supposed to be known functions. The criterion for choosing the ψ_{α}-functions and the approximation order N is based on the Bubnov-Galerkin principle: the best approximation is obtained if the functions obey the condition

$$\langle \psi_{\alpha}|H - E|\Psi^N \rangle = 0 \ . \tag{4.8}$$

The condition (4.8) is said to be the requirement of orthogonality for the discrepancy between $(H - E)\Psi^N$ and ψ_{α}-eigenfunctions. Let us take $|\Phi_0\rangle$ and $|\psi(E)\rangle$-functions as ψ_{α}.

If Ψ^N is chosen in the form of (4.6), then from the Bubnov-Galerkin principle it follows that

$$\left\langle \Phi_0|H - E| \left[a(E)\Phi_0 + \int dE' b(E')\psi(E') \right] \right\rangle = 0 \ ,$$

$$\left\langle \psi(E'')|H - E| \left[a(E)\Phi_0 + \int dE' b(E')\psi(E') \right] \right\rangle = 0 \ , \tag{4.9}$$

Note that, in reality, $b(E') = b(E, E')$. Hence

$$a(E)E_0 + \int dE' b(E')V^*(E') = a(E)E \ , \tag{4.10}$$

$$a(E)V(E') + E'b(E') = Eb(E') \ .$$

Fano has pointed out [4.3] that if $E = E'$, from the second equation one may find an expression for $b(E')$ using the generalized function which is introduced similarly to the symbolic Sokhotsky identity

$$b(E') = aV(E') \left[P\frac{1}{E - E'} + Z(E)\delta(E - E') \right] . \qquad (4.11)$$

The quantity $Z(E)$ will be defined below.

It is known that for large r

$$\psi(E) \sim \sin[k(E)r + \delta_0(E)] , \qquad (4.12)$$

where $\delta_0(E)$ is a slowly varying phase. Then, since Φ_0 tends to zero at large r

$$\Psi(E) \sim \int dE' b(E')\Psi(E') \sim aV(E)$$

$$\times \left\{ P\int \frac{dE'}{E - E'} \sin[k(E')r + \delta_0(E')] + Z(E)\sin[k(E)r + \delta_0(E)] \right\}$$

$$= aV(E)\{-\pi\cos[k(E)r + \delta_0(E)] + Z(E)\sin[k(E)r + \delta_0(E)]\} . \qquad (4.13)$$

where the integral is estimated along the properly chosen contour by using the residue theorem. Using the well known trigonometric identity we have

$$\Psi(E) \sim \frac{aV(E)Z(E)}{\cos\delta_r(E)} \sin[kr + \delta_0(E) + \delta_r(E)] . \qquad (4.14)$$

Here $\pi\cot\delta_r(E) = -Z(E)$.

The significance of the result obtained is that the presence of the states which are described by $|\psi_0\rangle$ results, in spite of the fact that they vanish when $r \rightarrow \infty$, in the additional phase shift δ_r. Moreover, we shall show that the phase shift is by no means small but has resonance character. Indeed, combining the first equation in (4.10) with (4.11), we have

$$aE_0 + a\int dE'|V(E')|^2 \left[P\frac{1}{E - E'} + Z(E)\delta(E - E') \right] = Ea . \qquad (4.15)$$

Hence, equating coefficients of a, we write

$$E_0 + P\int \frac{dE'|V(E')|^2}{E - E'} + (E)|V(E)|^2 = E . \qquad (4.16)$$

This expression helps us to define $Z(E)$:

$$Z(E) = \frac{E - E_0 - \Delta(E)}{|V(E)|^2} \equiv \frac{E - E_r}{|V(E)|^2} ,$$

$$\Delta E = P\int \frac{dE'|V(E')|^2}{E - E'} . \qquad (4.17)$$

Consequently, $\delta_r(E)$ changes rapidly and at $E_r \approx E_0 + \Delta E$ it passes through $\pi/2$, i.e., shows the resonance behaviour. The resonance width is given, according to (2.57), by the expression

$$\Gamma = -2(E - E_r) \tan \delta_r(E) = 2\pi |V(E)|^2 . \qquad (4.18)$$

Thus, we have demonstrated here possible resonance generation due to the interaction of discrete and continuous spectra. As is seen from (4.12), the resonance arises only when the energy of a discrete level reaches the continuous spectrum. In electron-atom collisions this occurs, for example, when the doubly exited negative-ion states are formed. The methods of calculation of the functions Φ_0 for those AIS and functions ψ for the continuous spectrum [do not confuse them with wave functions Ψ of the total Hamiltonian system, see (4.6)] should be given in order to obtain the final results (i.e., to find δ_0, E_r and δ_r). The solution of this problem will be obtained in subsequent sections.

Gailitis and *Damburg* [4.4] have considered a specific case of the appearance of such a resonance. First of all they noted that the analysis of the system leads to the following conclusion: In the asymptotic region ($r \gg a_0$), the equations which describe the inelastic electron scattering by the hydrogen atom (Chap. 3) can be written as the system of equations

$$\left[\frac{d_2}{dr^2} - \frac{l_2(l_2 + 1) + \alpha}{r^2} + k^2 \right] F(r) = 0 , \quad r \gg a_0 , \qquad (4.19)$$

when $n > 1$ excitation channels are degenerate. Here $l_2(l_2 + 1)$ is a diagonal matrix, $F(r)$ is a multicomponent function which is equal, e.g., for $L = 0$, to

$$F(r) = \begin{pmatrix} F_{2s}(r) \\ F_{2p}(r) \end{pmatrix} .$$

The matrix α couples the degenerate states with the same principal quantum number n. The terms of α/r^2 type arise only when orbital quantum numbers of these states differ by unity (e.g., the 2s and 2p states of an H atom).

Equations (4.19) may be transformed to the following form:

$$\left[\frac{d^2}{dr^2} - \frac{\lambda(\lambda + 1)}{r^2} + k^2 \right] A^{-1} f(r) = 0 , \qquad (4.20)$$

where $\lambda(\lambda + 1)$ is a diagonal matrix given by

$$A^{-1} [l_2(l_2 + 1) + \alpha] A = \lambda(\lambda + 1) .$$

Introducing $\lambda(\lambda + 1) = a$, we obtain

$$\lambda = -\tfrac{1}{2} \pm \left(\tfrac{1}{4} + a \right)^{1/2} . \qquad (4.21)$$

As follows from (4.20), the threshold behaviour of the quantity $F(r)$ and, therefore, of the transition amplitude, f, will be of the form $\sim k^{\lambda+1/2}$. Thus, at real $(\lambda + 1/2)$ the partial-wave excitation cross section tends to zero if $k \to 0$. However, if the quantities a are negative, and if $a < -\frac{1}{4}$, then $\lambda + \frac{1}{2} = i\mu$, where μ is real. Hence

$$|k^{\lambda+1/2}|^2 = |k^{i\mu}|^2 = 1 , \tag{4.22}$$

The latter means that the excitation cross section remains finite in the vicinity of the threshold. On the other hand, it is known that the condition $a < -\frac{1}{4}$ means that the differential equation has an infinite set of bound states which change resonances when the coupling with open channels is switched on. In other words, the finiteness of the excitation cross section at the threshold indicates the presence of an infinite set of resonances in the elastic cross section below the inelastic threshold. Resonances of this type are called dipole resonances, since the bound states result from the interaction of r^{-2} type. Therefore, dipole resonances are among those resonances described by Feshbach, Fano, and others.

Consider electron scattering by a hydrogen atom. For 2s-2p channels at $l = 0$ we have the asymptotic form of coupled equations

$$\left(\frac{d^2}{dr^2} + k^2 \right) f_{2s} - \frac{6}{r^2} f_{2p} = 0 ,$$

$$\left(\frac{d^2}{dr^2} + k^2 - \frac{2}{r^2} \right) f_{2p} - \frac{6}{r^2} f_{2s} = 0 , \tag{4.23}$$

where the α-matrix has the form $\left(\begin{smallmatrix} 0 & 6 \\ 6 & 0 \end{smallmatrix} \right)$, which gives for λ : $\lambda(\lambda+1) = 1 \pm (37)^{1/2}$. One of these values allows us to obtain the complex value $\lambda = -\frac{1}{2} + i\mu$, where $\mu = \left[(37)^{1/2} - \frac{5}{4} \right]^{1/2}$. Gailitis and Damburg have concluded that below the 2s and 2p excitations thresholds resonances must appear in the elastic scattering. This assumption has been confirmed experimentally.

There exists one more type of singularity of the scattering cross section near the inelastic threshold, the so-called "Wigner cusps" discussed in detail in [4.5].

4.2 Resonances in the Close-Coupling Method

Electron collisions with atoms may result in various types of reactions which produce two or more fragments. Speaking generally, each fragment is characterized by a certain intrinsic state and by the energy of its relative motion. The reaction channel is defined by the state of each fragment. The energetically allowed channels are called the open channels; while those forbidden in energy are the closed channels.

From more detailed theoretical studies of electron-atom interactions, e.g., [4.1, 4, 6] it is known that in elastic scattering there may also appear resonances different from the shape resonances when the closed channels are included into the expansion (3.44). We will analyze more carefully the mechanism of production of such a resonance and its theoretical interpretation for the ease of one open and one closed channel. It is very important that the methods expounded below allow a dynamical description of such resonances, i.e., an expression of their parameters in terms of the initial Hamiltonian. Note that these methods are useful in the treatment of nuclear resonances too. It is quite possible that in the near future they will be found to be precisely the methods which will be successful in elementary particle theory, where the problem of the dynamical description of resonances which arise in the scattering of elementary particles is not yet solved. In the low-energy region, where the nonrelativistic description is correct, further understanding of this important part of elementary particle physics will most likely be achieved exactly along this line. However, the introduction of the concept of a composite (quark) structure of "elementary" particles will be indispensable.

The mechanism of resonance production may be understood, e.g., from the solution of the two-channel Schrödinger equation for an electron scattered by a hydrogen atom. We shall follow the ideas of [4.1]. We retain only two terms in the expansion (3.44)

$$\Psi(r_1, r_2) = \phi_0(r_1) F_0(r_2) + \phi_1(r_1) F_1(r_2) , \tag{4.24}$$

where $\phi_{0,1}$ are hydrogen functions which obey (3.4) with eigenvalues E_0 and E_1. For the sake of simplicity the exchange is not included here.

Let the ϕ_0-function describe the ground 1s-state of a hydrogen atom; and the ϕ_1-function, the excited 2s (or 2p)-state. On substituting (4.24) into (3.1) we obtain the system (3.60) that in our case takes the following form:

$$\begin{aligned}
\left(\nabla^2 + k_0^2 - V_{00}\right) F_0(r) &= V_{01} F_1(r) , \\
\left(\nabla_2^2 + k_1^2 - V_{11}\right) F_1(r) &= V_{10} F_0(r) ,
\end{aligned} \tag{4.25}$$

where $k_n^2 = 2(E - E_n)$; potentials V_{nm} are given by (3.55); $E = E_0 + E_k$; E_k is the incident-electron energy. Combining the energy relations we obtain

$$k_0^2 = 2E_k ; \quad k_1^2 = k_0^2 + 2(E - E_1) . \tag{4.26}$$

Consider the case when the incident-electron kinetic energy is such that $k_1^2 < 0$, i.e., the 2s (or 2p) channel is closed. We introduce the notation

$$k_1^2 = -\kappa^2 = k_0^2 - \lambda^2 , \tag{4.27}$$

where $\lambda^2 = 2(E_1 - E_0)$ is the excitation energy.

We expand $F_{0,1}$ into a series over partial waves. Following the derivation of (3.80) [see also (3.93)], we have

$$\left[\frac{d^2}{dr^2} + k_0^2 - V_{00}^{LM} - l_2 (l_2 + 1) /r^2\right] f_0 = V_{01}^{LM} f_1 , \tag{4.28}$$

$$\left[\frac{d^2}{dr^2} - \kappa^2 - V_{11}^{LM} - l_2 (l_2 + 1) /r^2\right] f_1 = V_{10}^{LM} f_0 , \tag{4.29}$$

where $f_\alpha \equiv f^{LM}_{nl_1l_2}$. To understand this we need a physical interpretation of (4.28) and (4.29): equation (4.28) describes a single electron in the continuous spectrum (a second electron is, in accordance with (4.24), in the bound state), thus the asymptotic solutions of the equation will be of the form of (2.33b)

$$f_0(r \to \infty) = \frac{i^{l_2}}{k_0} \sin\left(k_0 r - \frac{\pi l_2}{2}\right) + f_{l_2}(k_0)e^{ik_0 r} . \tag{4.30}$$

We assert that (4.28) describes the open channel. Equation (4.29) is interpreted as describing the capture of the first (incident) electron by the excited-atom field (the second electron is in an excited, say, 2s or 2p state). This happens when the potential V_{11} is attractive and has a sufficient strength. Then the channel is closed and the electron wave functions vanish as $r \to \infty$ [$f_1 \to \exp(-\kappa r)$]. The latter corresponds to the formation of a hydrogen negative ion in the $2s^2$ or, say, $2p^2$ state. In the case of electron scattering by positive ions we obtain doubly excited states of the corresponding atom. It looks as if an incident electron has a partial probability to scatter instantly, and a partial one to be captured temporarily, forming a compound state. Such states (as a rule, with quite short lifetimes) in atomic physics are called AIS, as we have mentioned above.

Thus, our task is reduced to the solution of the system (4.28, 29) with the boundary condition (4.30) imposed on f_0. The approximation (4.28, 29) is nothing but CCM. Following it Feshbach [4.1] we shall show how resonances arise within this method. Let us confine ourselves, for the sake of simplicity, to the case when $l = 0$. The equation

$$\left(\frac{d^2}{dr^2} - \kappa_n^2 - V_{11}\right) X_n(r) = 0 \tag{4.31}$$

has a set of stationary solutions of the form $X_n \sim \exp(-\kappa_n r)$ with eigenvalues κ_n. It is obvious that X_n are eigenfunctions of bound states of the incident electron in the field V_{11} of an excited state of the hydrogen atom. We will seek a solution to the inhomogeneous euqation (4.29) in the form of a series over the set of solutions of homogeneous equations (4.31), considering the right-hand side of it as known

$$f_1(r) = \sum_n C_n X_n(r) ; \quad V_{10}f_0(r) = \sum_n a_n X_n(r) . \tag{4.32}$$

Upon substituting (4.32) into (4.29) and using (4.31) we obtain

$$f_1(r) = \sum_n X_n(r) \int dr' X_n(r')V_{10}(r')f_0(r')/\left(\kappa_n^2 - \kappa^2\right) . \tag{4.33}$$

Consider the solution of (4.28). The solution of the homogeneous equation which corresponds to (4.28) has the asymptotic form

$$f_0^{\text{hom}}(r \to \infty) = \sin(k_0 r + \delta_0) . \tag{4.34}$$

The partial solution of the inhomogeneous equation (4.28) may be expressed in terms of the Green's function

$$G(r, r') = \begin{cases} -f^1(r)f^2(r'), & r' > r; \\ -f^1(r')f^2(r), & r' < r, \end{cases} \tag{4.35}$$

where $f^{1,2}$ are the partial solutions of the homogeneous equation that corresponds to the inhomogeneous equation (4.28) with the following asymptotics:

$$f^1(r \to \infty) = k_0^{-1} \sin(k_0 r + \delta_0); \tag{4.36}$$

$$f^2(r \to \infty) = \exp[i(k_0 r + \delta_0)]. \tag{4.37}$$

The partial solutions may be found by the numerical methods explained in Sect. 2.4. The general solution of (4.28) is

$$f_0(r) = f_0^{\text{hom}}(r) + \sum_n A_n \int_0^\infty dr' G(r, r') V_{01}(r') X_n(r'), \tag{4.38}$$

where

$$A_n = \int V_{01}(r') f_0(r') X_n(r') dr' \left(\kappa_n^2 - \kappa^2\right)^{-1} \tag{4.39}$$

is an unknown constant.

In the sum over n in (4.38) we retain only one term for which $\kappa_n^2 \approx \kappa^2$. Then A_n may be determined upon substituting (4.38) into (4.39)

$$A_n = \frac{\int_0^\infty dr' X_n(r') V_{10}(r') f_0^{\text{hom}}(r')}{\kappa_n^2 - \kappa^2 - \Delta\kappa_n^2}, \tag{4.40}$$

where

$$\Delta\kappa_n^2 = \int_0^\infty dr' X_n(r') V_{10}(r') \int_0^\infty dr'' G(r', r'') V_{10}(r'') X_n(r''), $$

Consider now the asymptotics of both sides of (4.38). The left-hand side asymptotics has the form of (2.33b). Let's use the explicit form of the Green's function in the right-hand side, which gives

$$e^{-i\delta_0} \left(\sin k_0 r + f_0(k_0) e^{ik_0 r}\right) = \sin(k_0 r + \delta_0) + A_n \gamma^{1/2} e^{i(k_0 r + \delta_0)}, \tag{4.41}$$

where

$$\gamma^{1/2} = -\int_0^\infty f_0^{\text{hom}} V_{01} X_n dr. \tag{4.42}$$

On equating the coefficients of the same $\exp(ik_0 r)$ we obtain

$$f_0(k_0) = \frac{e^{2i\delta_0} - 1}{2ik_0} + \frac{e^{2i\delta_0}\gamma^{1/2} A_n}{k_0}. \tag{4.43}$$

Separating the integral into two terms $\Delta\kappa^2 = \mathrm{Re}\,\Delta\kappa^2 + i\gamma$ we represent (4.43) as

$$f_0(k_0) = \frac{e^{2i\delta_0} - 1}{2ik_0} + \frac{\gamma\,e^{2i\delta_0}}{k_0(\kappa_n'^2 - \kappa^2 - i\gamma)}\,, \qquad (4.44)$$

where

$$\kappa_n'^2 = \kappa_n^2 - \mathrm{Re}\,\Delta\kappa_n^2\,.$$

The first term in (4.44) describes the so-called potential scattering. The determination of the phase by the numerical solution of a homogeneous equation of type (4.28) is described in Sect. 2.4. The second addend is due to the closed channel and leads to a resonance of the Breit-Wigner type in the amplitude $f_0(k)$. The parameters of this resonance are calculated by using (4.40, 42) and determining eigenvalues of (4.31). Thus, we obtain the dynamical description of the resonance scattering within the chosen model. Resonances of this type are called Feshbach resonances or closed-channel resonances. These resonances are initiated by the interference between the continuous spectrum and doubly excited states.

The method described above is one of several possible realizations of the CCM. Its disadvantage is the necessity of solving equations both for continuous and discrete spectra. In addition, the answer is expressed by irregular solutions, whose numerical calculation is rather complicated. Moreover, the real resonance scattering cannot be calculated easily, unfortunately, within the above mentioned version of the CCM. First of all it must be supplemented by corresponding exchange terms [compare with (3.93)], which complicates the task enormously since differential equations are transformed into integro-differential equations with all the consequences. Moreover, in a realistic case one cannot confine onself to an s-wave, while inclusion of higher waves results in further complicating the system of equations.

If only one term is retained in the sum (3.44), this is an additional assumption which is not introduced in the full variant of CCM. Because of this assumption, it is impossible to derive a simple analytic expression for the scattering amplitude (4.44) in the Breit-Wigner form. Therefore, cumbersome numerical calculations must be performed. The restriction to a single term should be substantiated in each particular case.

And, finally, note that the system of close-coupled channel equations must be supplemented by, at least, both 2s and 2p excitation channels to describe the whole array of resonances below the $n = 2$ threshold, in accordance with results obtained by *Gailitis* and *Damburg* [4.4], and with the scope of CCM in general.

G. Schulz [4.7] was the first to conduct experimental studies on resonances and to discover the resonance at 19.3 eV in the elastic electron scattering by the He atom, which lies about 0.5 eV below the 2^3S threshold. Investigators have assigned this resonance to a quasi-bound state of the He^- negative ion with $1s2s^2$ configuration. This resonance is a typical closed-channel resonance. Similar resonances were discovered in inelastic e + He scattering, i.e., at $n = 2$

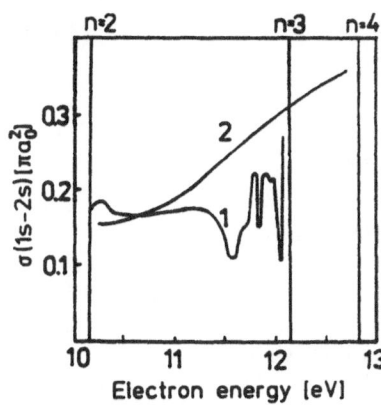

Fig. 4.1. Resonance behaviour of cross section for electron-impact excitation of the 2s level of the H atom. *1* – CMM calculation including the 1s, 2s, 2p states; *2* – CMM calculation including the 1s, 2s, 2p, 3s, 3p, 3d states

(19.95, 20.45, 21.00 eV). Not long after the ^{1}S-resonance was discovered in the elastic e + H scattering, *McGowan* and co-workers [4.8] carried out a series of precise experiments. A great number of theoretical calculations within the framework of the closed-coupled-channel scheme were performed too. *Burke* and *Taylor* [4.9] did vast close-coupling calculations which became a basis of the present-day CCM. The calculations within the 1s-2s-2p approximation gave an energy value of 9.56 eV at an energy width of 0.0475 eV. Later on, the p- and d-wave resonances were discovered below the $n = 3$ threshold (12.10 eV) [4.10]. The CMM calculations of these resonances were performed in [4.6]. Figure 4.1 shows the calculated 2s-excitation cross section for the hydrogen atom where the resonance area is presented on an enlarged (in comparison with Fig. 3.1) energy scale. The detailed discussion of this subject is given in Chap. 7.

In general, the complete CCM is very labour consuming when applied to scattering on a target more complicated than a hydrogen atom is considered. The diagonalization method to be described below is a simplified version of the CCM.

4.3 Autoionizing States of Two-Electron Systems

Thus, we have shown that an important role in the electron scattering by atoms and ions is played by the doubly excited states of the target, the total energy of which lies in a continuous spectrum of the corresponding atom or ion. As we have mentioned above (Sect. 4.1), in atomic physics these states are called the autoionizing states. It is necessary when calculating them to take the inter-electron interaction as accurately as possible into account. Before studying AIS we shall first examine the way the inter-electron interaction should be taken into account, with an example of the unexcited (ground) state of the helium atom.

The helium atom has a $1s^2$ configuration in its ground state. In this case the coordinate part of the wave function $\Psi(r_1, r_2)$ is symmetric and obeys the Schrödinger equation:

$$(H_0 + V)\Psi(r_1, r_2) = E\Psi(r_1, r_2) . \qquad (4.45)$$

Here

$$H_0 = -\frac{1}{2}\left(\nabla_1^2 + \nabla_2^2\right) - Z\left(\frac{1}{r_1} + \frac{1}{r_2}\right) ; \quad Z = 2 ; \quad V = \frac{1}{|r_1 - r_2|} .$$

In the zeroth approximation, Ψ can be represented as

$$\Psi(r_1, r_2) = \phi_{1s}(r_1)\phi_{1s}(r_2) ,$$

where ϕ_{1s} is a hydrogen-like function which obeys the equation

$$\left(-\frac{1}{2}\nabla_1^2 - \frac{Z}{r_1}\right)\phi_{1s}(r_1) = \mathcal{E}_{1s}\phi_{1s}(r_1) . \qquad (4.46)$$

Recall that the solution of (4.46) generaly has the form

$$\phi_{nlm}(r) = R_{nl}(r)Y_{lm}(\Omega) , \qquad (4.47)$$

where

$$R_{nl}(r) = \frac{1}{(2l+1)!}\sqrt{\frac{(n+l)!}{(n-l-1)!2n}}\left(\frac{2Z}{n}\right)^{3/2} e^{-Zr/n}$$

$$\times \left(\frac{2Z}{n}\right)^l F\left(-n+l+1, 2l+2, \frac{2Zr}{n}\right) ; \quad \mathcal{E}_n = -\frac{Z^2}{2n^2} . \qquad (4.48)$$

There are lots of different ways to take into account the inter-electron interaction influence on the energy of the $1s^2$-state, for instance, by a perturbation method. Recently, however, another method (the CI method) which allows one to obtain more accurate results with relative simplicity, has become popular. This method utilizes a many-configuration wave function of the form

$$\Psi(r_1, r_2) = \sum_{nm} \beta_{nm}\Psi_{nm}(r_1, r_2) , \qquad (4.49)$$

instead of the zero-approximation single-configuration wave function. Here the indices n and m denote the state configuration and show that the Ψ_{nm}-function which describes the (n, m)-configuration of the atom within the independent-electron approximation is constructed from the hydrogen-like functions ϕ_n and ϕ_m.

We substitute (4.49) into (4.45) using (4.46). Then

$$\sum_{nm} \beta_{nm}\left(\mathcal{E}_n + \mathcal{E}_m + V\right)\Psi_{nm}(r_1, r_2) = E\sum_{nm}\beta_{nm}\Psi_{nm}(r_1, r_2) . \qquad (4.50)$$

Multiplying the left-hand side of (4.50) by $\Psi_{n'm'}^*(r_1, r_2)$ and by integrating it over r_1, r_2 we write, using the orthogonality condition for $\phi_m(r)$

$$\beta_{nm} (E - \mathcal{E}_m - \mathcal{E}_n) = \sum_{m'n'} \beta_{n'm'} \langle \Psi_{nm} | V | \Psi_{n'm'} \rangle .\tag{4.51}$$

If one retains a finite number N of configurations in the expansion of (4.49), then it is possible to determine, from (4.51), N energy values $E^1, \ldots, E^N$ and the corresponding set of coefficients $\beta_{nm}^1, \ldots, \beta_{nm}^N$. If in (4.49) we keep only the $1s^2$-configuration, then it follows from (4.51) that

$$E = 2\mathcal{E}_{1s} + \langle 1s^2 | V | 1s^2 \rangle ,\tag{4.52}$$

which corresponds to the first-order of perturbation theory. The computation by (4.52) using (4.48) gives $E = -75\,\text{eV}$. But a much more accurate result can be obtained if a larger two-electron-state basis is used in (4.49). If the basis is chosen to be the twenty lowest-lying $1sns$-states the numerical solution of the system (4.51) gives $E = -77.53\,\text{eV}$, while the experimental value is $E = -79.01\,\text{eV}$.

Thus, we see that even a part of CI significantly increases the accuracy of the calculation. At present more perfect methods of taking into account the CI are formulated. Note that the correct choice of the initial basis configurations allows one to obtain an energy of the $1s^2$-state for the He atom very close to the experimental value. Section 5.1 deals with this problem in more detail.

As for the AIS, the calculation of the energy of the $2s^2$-state for the He atom by (4.51) with 10 configurations included (for details see Sect. 5.1) gives the value $E_{2s^2} = -21.1\,\text{eV}$. Experiment gives the value $\mathcal{E} = -21.2\,\text{eV}$. These data, as well as those on other highly excited He states presented in Table 5.1 and Fig. 5.1, testify to the accuracy of the CI method. All the numbers are presented on a scale where the energy of the ground state of He serves as the zero point.

Similar calculations may be carried out for the negative H^- ion states. For instance, the calculation of the $2s^2$-state of H^- which includes ten neighbouring configurations gives the value of $E_{2s^2} = -4.04\,\text{eV}$.

Since the energy of a single ionized He atom on this scale is equal to $-54.5\,\text{eV}$, the AIS may decay producing, for example, the He^+ ion in the $1s$-state and a free electron with the energy $(54.5 - 21.1) = 33.4\,\text{eV}$. This means that in the process of electron scattering by the He^+ ion at an incident-electron energy of $33.4\,\text{eV}$, the resonance generated by this AIS can be observed. This is precisely what we mean when we say that the AIS corresponds to the continuous He spectrum (i.e., to the singly ionized He spectrum).

As in the case mentioned above, the $2s^2$-state of the H^- negative ion may autoionize, which is accompanied by the transition of the first electron into the $1s$-state and by the transition of the second electron into the continuous spectrum with the energy $(-4.04 + 13.6) = 9.56\,\text{eV}$. More detailed information on the AIS of the H^- negative ion is presented in Sect. 7.2. The energies of other atoms and ions are calculated in a similar way. This problem will be discussed later.

The matrix elements entering (4.51) are frequently used in calculations involving electron-atom collisions. For complex atoms the computation of such matrix elements is very tedious. The method suggested by *Fano* [4.11] is discussed in detail in [4.12] (see also Ref. [4.13]).

Until now we have confined ourselves to those AIS which are formed in a two-electron excitation process. In the case of complex atoms the formation of AIS through one-electron excitation from an inner shell is also quite possible. However, a process of this kind is beyond the class of problems considered here.

Interest in AIS is due to their frequent occurrence in various atomic processes, namely, besides electron-atom collisions, in the collisions of atomic particles or in interactions of different particles with radiation. In electron-atom collisions, the AIS arise a resonances in elastic scattering as well as in excitation and ionization. Here we shall confine ourselves to collision processes in the incident-electron energy range below the ionization threshold, in other words, to the resonances in scattering processes.

Consider, for example, elastic scattering of electrons by a hydrogen atom. Let the incident-electron energy be less than 10.2 eV, i.e., excitation does not occur. It turns out that a resonance is observed at 9.56 eV. As mentioned the reason is that at the incident energy $\mathcal{E} = 9.56$ eV the total energy of the system "unexcited atom + electron" has a value of $E = -4.04$ eV (i.e., the energy of the AIS (Sect. 7.2). This means that, besides the direct electron scattering, a temporary electron capture by the atom occurs which is related to formation of the negative ion. In other words, at this energy the number of scattered electrons is changed significantly since the electron receives an additional probability of being scattered. The interference between the direct scattering amplitude and that of the scattering via the formation of AIS produces the structure in the cross section in the form of a maximum (constructive interference) or a minimum (destructive one).

This resonance is sometimes also considered as a process of incident-electron capture when the captured electron orbits for a certain time around the nucleus and then leaves the atom. If one considers the evolution of the wave function in time, then the increase of a cross section due to the formation of AIS can be explained as follows. As has been mentioned in Chap. 2, the cross section is the ratio of the scattered particle flux to that of the incident particles. Since the incident electron might be captured by the atom and emitted only some time later, the situation may arise that the incident flux is over while the electron ejection still continues. In this case the cross section grows abruptly. It should be noted, however, that this time delay can also cause a decrease in the cross section since at a given incident flux the scattered flux may be reduce sharply.

The question about the origin of spontaneous decay of AIS is very important. An AIS may decay both without emission (due to the inter-electron interaction) and by way of photon emission. In the case of multicharged ions the lines which correspond to this decay are called satellites. For example, the transition from the $2p^2$ autoionizing state of a multicharged helium-like iron ion into the state with a 1s2p configuration produces a satellite line close to the resonance transition line 2p-1s in the hydrogen-like iron ion. the first process is usually predominant in the low-energy region for neutral atoms and low-charged ions except in cases where the nonradiative transition is forbidden.

5. Diagonalization Method

Here we present a simplified version of Feshbach's method, the diagonalization method (DM), which we apply hereafter to resonances in electron-atom and electron-ion scattering. The methods of calculation of the parameters of AIS are also discussed in the present chapter. An application of the DM to the elastic scattering of an electron by an He$^+$ ion is considered in detail. In order to elucidate the essence of the problem we shall confine ourselves to the case when only the channel of elastic scattering is open.

5.1 Basic Ansatz

Its complexity is a serious disadvantage of the CCM. In practice, the inclusion of two channels is insufficient. But the larger the number of channels, the higher the number of equations is and, hence, the increased computational difficulties. The diagonalization method was proposed by *Balashov* et al. [5.1–3] as a simplified version of Feshbach's method [5.4].

This method includes a sufficient number of closed channels corresponding to different configurations (i.e., to excited states of a two-electron system) and at the same time does not increase the number of differential equations to be solved. Moreover, sometimes this method allows one to take into account the inter-configurational interaction more exactly than the CCM does.

The DM has been used successfully for electron-impact ionization and photoionization of atoms. In the scattering problem the DM was first utilized in [5.5]. The method requires some modification and complication in comparison with the case considered in [5.2].

Thus, one again needs to solve the Schrödinger equation. Instead of looking for an approximate solution in the form of expansion (3.44) which comprises the terms corresponding both to the closed and open channels, we shall retain in this expansion only the terms describing the open channels, whereas the closed channels will be described by a linear combination of $\Phi_\mu(r_1, r_2)$-functions characterizing the AIS and obtained by the diagonalization of the Hamiltonian on the subspace of closed channels.

To solve (4.45), let us represent $\Psi(r_1, r_2)$ by the following expansion:

$$\Psi(r_1, r_2) = \sum_{n=1}^{N} A\left[\phi_n(r_1)\widetilde{F}_n(r_2)\right] + \sum_{n>N} A\left[\phi_n(r_1)\widetilde{F}_n(r_2)\right]$$
$$+ \int d\alpha A[\phi_\alpha(r_1)F_\alpha(r_2)] , \qquad (5.1)$$

where A is an antisymmetrization operator, N is the number of open channels and $\phi_n(r)$ and $\phi_\alpha(r)$ are the eigenfunctions of the discrete and continuous spectra of the target, e.g., the solutions of (4.46). Neglecting in (5.1), as well as in CCM, the integral over the continuous spectrum and representing $\widetilde{F}_n(r)$ for $n > N$ in the form

$$\widetilde{F}_n(r) = \sum_{m=1} C_{mn}\phi_m(r) , \qquad (5.2)$$

we shall rewrite the expansion (5.1) in the following form:

$$\Psi(r_1, r_2) = \sum_{n \leq N} A[\phi_n(r_1)F_n(r_2)] + \sum_{n,m>N} C_{nm} A[\phi_n(r_1)\phi_m(r_2)] , \qquad (5.3)$$

where

$$F_n(r_2) = \widetilde{F}_n(r_2) + \sum_{m \leq N} C_{mn}\phi_m(r_2) . \qquad (5.4)$$

Note that the wave function (5.3) is an approximation of an exact wave function of the system. This results from the neglect of the integral over the continuous spectrum in (5.1) and (5.2). Here and below, we must pass from the set of antisymmetrized functions $X_{mn}(r_1, r_2) = A[\phi_n\phi_m]$ to that of other antisymmetrized functions $\Phi_{nm}(r_1, r_2)$

$$\Phi_{mn}(r_1, r_2) = \sum_{m'n'} \beta_{mn,m'n'} X_{m'n'}(r_1, r_2) . \qquad (5.5)$$

Inversion of (5.5) produces the expression for $X_{mn}(r_1, r_2)$

$$X_{mn}(r_1, r_2) = \sum_{m'n'} \beta^{-1}_{mn,m'n'} \Phi_{m'n'}(r_1, r_2) . \qquad (5.6)$$

Here the coefficients $\beta_{mn,m'n'}$ determine the configuration mixing and are found from the condition that they form an orthonormal set and diagonalize the total Hamiltonian H on the subspace of the functions $X_{mn}(r_1, r_2)$

$$\langle \Phi_{m'n'}(r_1, r_2)|H|\Phi_{mn}(r_1, r_2)\rangle = \mathcal{E}_{mn}\delta_{mn,m'n'} . \qquad (5.7)$$

Let us lable, for convenience, the sequence mn by a single number $\mu = 1, 2, \ldots$. Then (5.5–7) become

$$\Phi_\mu(r_1, r_2) = \sum_\nu \beta_{\mu\nu} X_\nu(r_1, r_2) ; \qquad (5.8)$$

$$\langle \Phi_\nu(r_1, r_2)|H|\Phi_\mu(r_1, r_2)\rangle = \mathcal{E}_\mu \delta_{\mu\nu} ; \tag{5.9}$$

$$X_\mu(r_1, r_2) = \sum_\nu \beta_{\mu\nu}^{-1} \Phi_\nu(r_1, r_2) , \tag{5.10}$$

where $\mathcal{E}_\mu$ is the energy of the system in the Φ_μ state. Substitution of (5.10) into (5.3) gives

$$\Psi(r_1, r_2) = \sum_{n=1}^{N} A[\phi_n(r_1)F_n(r_2)] + \sum_\mu \Lambda_\mu \Phi_\mu(r_1, r_2) , \tag{5.11}$$

where

$$\Lambda_\mu = \sum_\nu \beta_{\mu\nu}^{-1} C_\nu .$$

In the total spin representation, there exist two wave functions $\Psi(r_1, r_2)$, namely $\Psi^{S=0}(r_1, r_2)$ and $\Psi^{S=1}(r_1, r_2)$, corresponding to the singlet and triplet states of the system "electron + helium ion", respectively. Taking this into account, we have from (5.11)

$$\Psi^S(r_1, r_2) = \frac{1}{(2)^{1/2}} \sum_{n=1}^{N} [\phi_n(r_1)F_n^S(r_2) + (-1)^S \phi_n(r_2)F_n(r_1)]$$
$$+ \sum_\mu \Lambda_\mu^S \Phi_\mu^S(r_1, r_2) . \tag{5.12}$$

The wave functions $\Phi_\mu^S(r_1, r_2)$ describe the discrete states of the system which are embedded in the continuum (i.e., the AIS of the He atom). The expansion of the total wave function $\Psi^S(r_1, r_2)$ of the system in the form of (5.12) was used earlier by *Gailitis* [5.6] and *Burke* [5.7] with variational functions $\Phi_\mu^S(r_1, r_2)$. It should also be noted that the expansion (5.12) has an important extremum property [5.6]: if all open channels at a given energy in the system are contained in the first sum in (5.12), then the scattering eigenphases tend to their exact values from below with the increasing of number of functions Φ_μ. Let us note that the functions $F_n^S(r)$ and coefficients Λ_μ^S in (5.12) are unknown quantities.

Consider, for simplicity, the elastic electron scattering by a hydrogen atom below the excitation threshold without accounting for exchange. We shall look for a solution Ψ which is characterized both by the total spin S and the orbital angular momentum L (and, of course, by their projections), as well as by parity π. Due to the presence of a single open channel, the expression (5.11) assumes the form

$$\Psi^{LS\pi}(r_1, r_2) = F_0^{LS\pi}(r_1)\phi_0(r_2) + \sum_\mu \Lambda_\mu^{LS\pi} \Phi_\mu^{LS\pi}(r_1, r_2) . \tag{5.13}$$

Here $\phi_0 = \phi_{100}$ is a wave function of the ground state H atom.

The orthonormalized wave functions of AIS have, in accordance with (5.5), the following form in the total angular momentum presentation:

$$\Phi_\mu^{LS\pi}(r_1, r_2) = \sum_{n_1 l_1 n_2 l_2} \beta_\mu(n_1 l_1, n_2 l_2) B_{n_1 l_1 n_2 l_2}^{LM}(r_1, r_2) \qquad (5.14)$$

where the basis functions are

$$B_{n_1 l_1 n_2 l_2}^{LM}(r_1, r_2) = \sum_{m_1 m_2} C_{l_1 m_1 l_2 m_2}^{LM} \phi_{n_1 l_1 m_1}(r_1) \phi_{n_2 l_2 m_2}(r_2) .$$

Here C^{LM} are Clebsch-Gordan coefficients, ϕ_{nlm} are hydrogen wave functions with eigenvalues of $\mathcal{E}_{nl}^0$, and $n_1 \geq 2$, $n_2 \geq 2$ in (5.14).

The coefficients β_μ only characterize the CI (note that if an AIS is formed due to the capture of an incident electron which moves along the Z-axis, then $M = 0$). The coefficients β_μ are found from the condition that Φ_μ forms an orthonormalized system and diagonalizes the Hamiltonian in accordance with (5.9).

It is easy to show that in this case the following system of equations is obtained from (5.8,9):

$$\langle B_{n_1 l_1 n_2 l_2}^{LM} | H - \mathcal{E}_\mu | \Phi_\mu \rangle = 0 . \qquad (5.15)$$

Since the functions B are orthonormalized, the substitution of (5.14) into (5.15) gives

$$\beta_\mu(n_1 l_1, n_2 l_2) \left(\mathcal{E}_{n_1 l_1}^0 + \mathcal{E}_{n_2 l_2}^0 - \mathcal{E}_\mu \right)$$

$$= - \iint dr_1 dr_2 B_{n_1 l_1 n_2 l_2}^{LM*}(r_1, r_2) V_{12}(r_1, r_2)$$

$$\times \sum_{n_1' l_1' n_2' l_2'} \beta_\mu(n_1' l_1', n_2' l_2') B_{n_1' l_1' n_2' l_2'}^{LM}(r_1, r_2) . \qquad (5.16)$$

The system (5.16) of algebraic equations is nothing but a detailed presentation of the system (4.51). While carrying out practical calculations one may confine oneself to a finite basis of the configurations in (5.14). The solution of the system gives quite exact values for the energies of the AIS, $\mathcal{E}_\mu$, and the coefficients β_μ. In other words, it allows one to calculate the energies and wave functions of an AIS without resorting to a perturbation method. The obtained accuracy depends on the choice of the initial basis.

We have examined the way in which the energies and wave functions of an AIS should be sought on an example of the H$^-$ ion. The same CI method may be used for more complex atoms and ions too. Since $n_1 \geq 2$ and $n_2 \geq 2$ in (5.14), the functions Φ_μ are orthogonal to the first term in (5.13). The F_0-function contains all the information about the scattering, to find this function is our aim. Substituting (5.14) into (4.45) and applying (4.46) we have

$$\left(-\tfrac{1}{2}\Delta_1 - (E - E_0) + V_1 + V_{12}\right) F_0(r_1)\phi_0(r_2) + \sum_\mu \Lambda_\mu(H - E)\Phi_\mu = 0 ,$$

$$V_1 = -1/r_1 , \quad V_{12} = \frac{1}{|r_1 - r_2|} . \qquad (5.17)$$

(Here and to the end of this section we shall omit the superscript $\{LS\pi\}$.) Then, multiplying the left-hand side of (5.17) by ϕ_0^* and integrating it over r_2, we shall write, taking into account the orthogonality of the functions,

$$-\tfrac{1}{2}\Delta_1 F_0 - (E - E_0)F_0 + (\tilde{V}_1 + \tilde{V}_{12})F_0 + \sum_\mu \Lambda_\mu \langle \phi_0|H|\Phi_\mu \rangle = 0 , \qquad (5.18)$$

where

$$\tilde{V}_1 + \tilde{V}_{12} = \langle \phi_0|V_1 + V_{12}|\phi_0 \rangle = -\frac{1}{r_1} + \langle \phi_0|V_{12}|\phi_0 \rangle . \qquad (5.19)$$

Using (4.46) we obtain

$$\langle \phi_0|H|\Phi_\mu \rangle = \langle \phi_0|V_{12}|\Phi_\mu \rangle , \qquad (5.20)$$

since the orthogonality condition $\langle \phi_0|\Phi_\mu \rangle = 0$ is valid. Then, with (5.19) taken into account, (5.18) assumes the form

$$\Delta_1 F_0(r_1) + k_0^2 F_0(r_1) - \left(\tilde{U}_1 + \tilde{U}_{12} \right) F_0(r_1) = \sum_\mu \Lambda_\mu \langle \phi_0|U_{12}|\Phi_\mu \rangle , \qquad (5.21)$$

where

$$k_0^2 = 2(E - E_0) , \quad U = 2V .$$

Multiplying the left-hand side of (5.17) by Φ_ν^* and integrating it over r_1, r_2, we write

$$\sum_\mu \Lambda_\mu \langle \Phi_\nu|H|\Phi_\mu \rangle + \langle \Phi_\nu|U_{12}|\phi_0 F_0 \rangle = E\Lambda_\nu . \qquad (5.22)$$

Hence, using (5.6),

$$\Lambda_\mu = \frac{\langle \Phi_\mu|U_{12}|\phi_0 F_0 \rangle}{E - \mathcal{E}_\mu} . \qquad (5.23)$$

The value of Λ_μ is still unknown since the solution of (5.21) is unknown. Let us now solve (5.21) by using the Green's function for the corresponding homogeneous equation. As earlier, we have

$$F_0(r_1) = F_{0\,\text{hom}}^{(+)}(r_1) + \int dr_1' G(r_1, r_1') \sum_\mu \Lambda_\mu \langle \phi_0(r_2')|U_{12}|\Phi_\mu(r_1', r_2') \rangle . \, (5.24)$$

The substitution of F_0 into (5.23) produces an algebraic equation for Λ_μ

$$\Lambda_\mu(E - \mathcal{E}_\mu) = \langle \Phi_\mu(r_1', r_2')|U_{12}(r_1', r_2')|\phi_0(r_2')F_{0\,\text{hom}}^{(+)}(r_1') \rangle$$
$$+ \sum_\nu \int \Lambda_\nu \langle \Phi_\nu(r_1', r_2')|U_{12}(r_1', r_2')|\phi_0(r_2') \rangle$$
$$\times \int dr_1'' G(r_1', r_2'') \langle \phi_0(r_2''|U_{12}(r_1'', r_2'')|\Phi_\mu(r_1'', r_2'') \rangle dr_1' . \qquad (5.25)$$

The solution of (5.21) without the right-hand side, $F_{0\,\text{hom}}(r_1)$, is a linear combination of two solutions which we shall denote by $F_{0\,\text{hom}}^{(\pm)}(r_1)$, where $F_{0\,\text{hom}}^{(+)}(r_1)$ is a regular solution, i.e., $F_{0\,\text{hom}}^{(+)}(0) = 0$. The Green's function $G(r, r')$ is a solution of the following equation:

$$\left[\Delta + k_n^2 - (U_1 + U_{12}) \right] G(r, r') = \delta(r - r') \tag{5.26}$$

and may be written

$$G = \frac{1}{k_n^2 - H_0 + i\varepsilon} , \tag{5.27}$$

where $H_0 = -\Delta + U_1 + U_{12}$. We take advantage of the asymptotic form of the Green's function [5.8]

$$G(r, r') \underset{r \simeq \infty}{\sim} 2 \frac{e^{ik_0 r}}{r} F_{0\,\text{hom}}^{(+)}(r') . \tag{5.28}$$

Applying the symbolic Sokhotsky identity to (5.27) we have:

$$G = \frac{P}{k_0^2 - H_0} - i\pi\delta(k_0^2 - H_0) . \tag{5.29}$$

Here the δ-function is the following operator:

$$\delta(k_0^2 - H_0) = |F_{0\,\text{hom}}^{(+)}\rangle\langle F_{0\,\text{hom}}^{(-)}| .$$

Using (5.28) we obtain from (5.24) the scattering amplitude:

$$f = f_{\text{pot}} + 2 \sum_\mu \Lambda_\mu \langle F_{0\,\text{hom}}^{(+)}(r_1)\phi_0(r_2)|V_{12}|\Phi_\mu(r_1, r_2)\rangle , \tag{5.30}$$

where f_{pot} is a potential scattering amplitude.

The value of Λ_μ must be found by solving the system of equations (5.25). If elements of the system are represented in matrix form, then the terms Λ_μ will be placed along the diagonal. Balashov and co-workers have given the arguments in favor of the assumption that one may confine oneself to the diagonal elements in (5.25) without an appreciable loss of accuracy [5.2]. It was precisely this Ansatz that was the reason of introducing the term "diagonalization method".

Thus, we shall confine ourselves to the terms $\mu = \nu$ in the sum over ν in (5.25). Using now an explicit expression of the Green's function in the form of (5.29) we obtain:

$$\Lambda_\mu = \frac{\langle \Phi_\mu(r_1, r_2)|V_{12}|F_{0\,\text{hom}}^{(+)}(r_1)\phi_0(r_2)\rangle}{E - \mathcal{E}_\mu - \Delta_\mu + i\Gamma_\mu/2} \tag{5.31}$$

where:

$$\Delta_\mu = \left\langle \Phi_\mu|V_{12}\phi_0 \frac{P}{k_0^2 - H_0}\phi_0 V_{12}|\Phi_\mu \right\rangle , \tag{5.32}$$

$$\Gamma_\mu = 2\langle \Phi_\mu|V_{12}|F_{0\,\text{hom}}^{(+)}\phi_0\rangle\langle\phi_0 F_{0\,\text{hom}}^{(+)}|V_{12}|\Phi_\mu\rangle . \tag{5.33}$$

The quantity Γ_μ may be represented as:

$$\Gamma_\mu = 2|\langle \Phi_\mu | V_{12} | F_{0\,\mathrm{hom}}^{(+)} \phi_0 \rangle|^2 \qquad (5.34)$$

and it is a resonance width. If only one term is kept in the sum over μ in (5.30), then the scattering amplitude is:

$$f = f_{\mathrm{pot}} + \frac{\pi \Gamma}{E - \mathcal{E}_\mu - \Delta_\mu + i\Gamma/2} \ . \qquad (5.35)$$

Thus, as has been noted in the CCM, the scattering amplitude is expressed through two terms, one of which has a resonance form.

The representation (5.30) is not, however, used in practice because it is impossible to perform certain calculations in three dimensions. In order to do that, one needs to consider a radial equation in (5.21).

5.2 Radial Equations

As mentioned above, the transition to the radial equations it is also necessary to carry out certain calculations within the framework of the DM. For this one needs to expand the function $F_0(r_1)$ in a series over partial waves as given by (3.65) with $l = L$, $m = M$, since in this case the target is in a state with zero momentum. We rewrite (5.21) in the following way:

$$\left(\Delta_1 + k_0^2 + \frac{2Z}{r_1} - \tilde{U}_{12} \right) F_0(r_1) = \sum_\mu \Lambda_\mu \langle \phi_0 | U_{12} | \Phi_\mu \rangle \ . \qquad (5.36)$$

Substituting the partial-wave expansion of F_0 in (5.36), we obtain

$$\sum_L \frac{\left[\Delta_1 + k_0^2 + 2Z/r_1 - U_{12} \right] f_{0L}(r_1)}{r_1} Y_{LM}(\Omega_1) = \sum_\mu \Lambda_\mu \langle \phi_0 | U_{12} | \Phi_\mu \rangle \ . (5.37)$$

Multiplying (5.37) by $Y_{L'M'}^*(\Omega_1)$ and integrating the product over Ω_1, we have

$$\left(\frac{d^2}{dr_1^2} + k_0^2 + \frac{2Z}{r_1} - \frac{L'(L'+1)}{r_1^2} - U_{12} \right) f_{0L'}(r_1) = \sum_\mu \Lambda_\mu \chi_\mu^{L'M'}(r_1) \ , \quad (5.38)$$

where

$$\chi_\mu^{LM}(r_1) = r_1 \int d\Omega_1 Y_{LM}^*(\Omega_1) \langle \phi_0(r_2) | U_{12} | \Phi_\mu(r_1, r_2) \rangle \ .$$

Consider the right-hand side of (5.38) in more detail. By definition:

$$\langle \phi_0 | U_{12} | \Phi_\mu \rangle = \int dr_2 Y_{00}(\Omega_2) R_{10}(r_2) \frac{1}{|r_1 - r_2|}$$

$$\times \sum_{n_1 l_1 n_2 l_2} \beta_\mu(n_1 l_1 n_2 l_2) \sum_{m_1 m_2} C^{LM}_{l_1 m_1 l_2 m_2} \phi_{n_1 l_1 m_1}(r_1)$$

$$\times \phi_{n_2 l_2 m_2}(r_2) . \tag{5.39}$$

Here (5.14, 17) are taken into account. If we recall (4.47) and expand the potential in a series over spherical functions, then

$$\chi^{L'M'}_\mu(r_1) = \sum_{\substack{n_1 l_1 m_1 k \\ n_2 l_2 m_2 q}} r_1 \int r_2^2 dr_2 \int d\Omega_1 \int d\Omega_2 Y^*_{L'M'}(\Omega_1)$$

$$\times Y_{l_1 m_1}(\Omega_1) Y_{kq}(\Omega_1) Y_{00}(\Omega_2) Y_{l_2 m_2}(\Omega_2) Y^*_{kq}(\Omega_2)$$

$$\times R_{10}(r_2) \gamma_k \beta_\mu C^{LM}_{l_1 m_1 l_2 m_2} R_{n_1 l_1}(r_1) R_{n_2 l_2}(r_2) . \tag{5.40}$$

Using the properties of the spherical functions and of the $3j$-symbols we may write (5.40) as

$$\chi^{L'M'}_\mu(r_1) = \sum_{\substack{n_1 l_1 m_1 \\ n_2 l_2 m_2}} \sum_k r_1 \int r_2^2 dr_2 \beta_\mu C^{LM}_{l_1 m_1 l_2 m_2} (-1)^{M'}$$

$$\times ([L'][l_1][l_2])^{1/2} \begin{pmatrix} L' & l_1 & l_2 \\ 0 & 0 & 0 \end{pmatrix} \begin{pmatrix} L' & l_1 & l_2 \\ -M' & m_1 & m_2 \end{pmatrix}$$

$$\times \gamma_k R_{10}(r_2) R_{n_1 l_1}(r_1) R_{n_2 l_2}(r_2) . \tag{5.41}$$

By definition

$$C^{LM}_{l_1 m_1 l_2 m_2} = (-1)^{l_1 + l_2 - M} (2L+1)^{1/2} \begin{pmatrix} l_1 & l_2 & L \\ m_1 & m_2 & -M \end{pmatrix} . \tag{5.42}$$

Thus:

$$\chi^{L'M'}_\mu = \sum_{\substack{n_1 l_1 m_1 \\ n_2 l_2 m_2}} \sum_k (-1)^{l_1 + l_2 - M + M'} ([L][L'][l_1][l_2])^{1/2}$$

$$\times \begin{pmatrix} l_1 & l_2 & L \\ m_1 & m_2 & -M \end{pmatrix} \begin{pmatrix} L' & l_1 & l_2 \\ 0 & 0 & 0 \end{pmatrix} \begin{pmatrix} L' & l_1 & l_2 \\ -M' & m_1 & m_2 \end{pmatrix}$$

$$\times r_1 \int r_2^2 dr_2 \gamma_k R_{10}(r_2) R_{n_1 l_1}(r_1) R_{n_2 l_2}(r_2) . \tag{5.43}$$

Hence:

$$\sum_{m_1 m_2} \begin{pmatrix} l_1 & l_2 & L \\ m_1 & m_2 & -M \end{pmatrix} \begin{pmatrix} L' & l_1 & l_2 \\ -M' & m_1 & m_2 \end{pmatrix} = \frac{\delta_{LL'} \delta_{MM'}}{2L+1} . \tag{5.44}$$

Then

$$\chi^{L'M'}_\mu(r_1) = \sum_{n_1 l_1 n_2 l_2} (-1)^{l_1 + l_2} ([l_1][l_2])^{1/2} \begin{pmatrix} L' & l_1 & l_2 \\ 0 & 0 & 0 \end{pmatrix}$$

$$\times \beta_\mu \delta_{LL'} \delta_{MM'} r_1 \int R_{n_1 l_1}(r_1) r_2^2 dr_2 \gamma_{l_2} R_{10}(r_2) R_{n_2 l_2}(r_2) . \tag{5.45}$$

The problem is therefore reduced to solving the following equation:

$$\left(\frac{d^2}{dr_1^2} + k_0^2 + \frac{2Z}{r_1} - \frac{L'(L'+1)}{r_1^2} - \tilde{U}_{12}\right) f_{0L'}(r_1) = \sum_\mu \Lambda_\mu \chi_\mu^{L'M'}(r_1) . \quad (5.46)$$

If one then substitutes the partial-wave expansion of F_0 into (5.23), Λ_μ is

$$\Lambda_\mu(E - \mathcal{E}_\mu) = \int dr_1 f_{0L}(r_1) \chi_\mu^{*LM}(r_1) . \quad (5.47)$$

Thus, the situation looks like a two-channel version of CCM; although Λ_μ in the right-hand side of (5.46) depends on f_{0L}, this dependence is in fact reduced to an unknown constant. If the right-hand side of (5.46) depended on $f_{0L}(r)$ explicitly, then the solution of this equation would be difficult. As regards our case, we shall do the same as we have done in the previus section, namely, we shall express the solution of (5.46) in terms of the Green's function:

$$f_{0L}(r) = f_{0L \, \text{hom}}(r) + \int dr' G(r,r') \sum_\mu \Lambda_\mu \chi_\mu^{LM}(r') . \quad (5.48)$$

This is the formal solution of the problem.

5.3 Example: Resonances in Elastic Electron Scattering by Helium Ions

As an example, we apply the method described above to the resonances produced in the elastic scattering of electrons by helium ions. Despite the fact that this process resembles the electron scattering by a hydrogen atom, which has been the topic of a number of papers, it has a number of significant differences. The structure of the electron scattering by a hydrogen atom is affected by excited states of the negative hydrogen ion, the experimental study of which is a quite complicated subject. The resonances in the process considered below are generated by autoionizing states of atomic helium. The experimental data on these states are abundant (in particular, those obtained in photoabsorption studies), unlike the previous case. The influence of the closed channels in this case is manifested more clearly. In general, the AIS of the neutral atoms are more pronounced than doubly excited negative ion states.

In the screened decreasing field of an excited atom (a Yukawa-type potential plus a polarization potential), a limited number of resonances is observed in the electron scattering by atoms. This pattern closely resembles that for nuclear particle interaction. The interaction potential for two nucleons (neutron and proton) in the deuteron also has the form of a Yukawa potential. Because of the rapid decrease of this potential, the deuteron has no excited states at all, i.e., the nuclear forces are actually quite weak. This is not valid for the hydrogen atom since its excited states have a potential form different from that of Yukawa's type, due to the degeneracy of the hydrogen atom states (Sect. 3.1).

An infinite sequence of the AIS below each threshold may occur in the Coulomb field of an excited He atom and thus, the same holds true for the infinite number of resonances in e + He$^+$ scattering. The Coulomb field requires a special consideration to be explained below.

The long-range Coulomb attraction which leads to the temporary ($\sim 10^{-14}$s) capture of an incident electron by an excited He$^+$ ion is the physical mechanism responsible for producing these resonances. After that time, the first electron moves away while the second returns to the 1s-shell due to the inter-electron repulsion.

The problem of electron elastic scattering by an He$^+$ ion requires solving the stationary Schrödinger equation (2.1). This is done by representing $\Psi(r_1, r_2)$ in the form

$$\Psi^S(r_1, r_2) = 2^{-1/2} \sum_\alpha \left[\phi_\alpha(r_1) F_\alpha^S(r_2) + (-1)^S \phi_\alpha(r_2) F_\alpha^S(r_1) \right] . \qquad (5.49)$$

The DM allows one to obtain the dynamical description of the resonances, i.e., to describe the observable parameters in terms of the matrix elements of the initial Hamiltonian.

Consider the case when a single (elastic) channel is open. Let us write the first term in (5.49) that describes the contribution of this channel explicitly. The closed-channel contributions are described by $\Phi_\mu^S(r_1, r_2)$-functions. Then, neglecting the integral over the continuous spectrum, we may express the wave function of the system by

$$\Psi^S(r_1, r_2) = 2^{-1/2} \left[\phi_0(r_1) F_0^S(r_2) + (-1)^S \phi_0(r_2) F_0^S(r_1) \right] + \sum_\mu \Lambda_\mu^S \Phi_\mu^S . (5.50)$$

Substituting (5.50) into (4.45) and using the Bubnov-Galerkin principle [5.9], we obtain the system of linear equations for the coefficients Λ_μ^S and $F_0^S(r)$-functions

$$\Lambda_\mu^S = \frac{2^{-1/2} \langle \Phi_\mu^S(r_1, r_2) | r_{12}^{-1} | \phi_0(r_1) F_0^S(r_2) + (-1)^S \phi_0(r_2) F_0^S(r_1) \rangle}{E - \mathcal{E}_\mu} , \qquad (5.51)$$

$$\left[-\frac{1}{2} \nabla_2^2 - \frac{2}{r_2} + \mathcal{E}_0 + \langle \phi_0(r_1) | r_{12}^{-1} | \phi_0(r_1) \rangle - E \right] F_0^S(r_2)$$
$$+ (-1)^S \langle \phi_0(r_1) | r_{12}^{-1} | F_0^S(r_1) \rangle \phi_0(r_2)$$
$$= 2^{1/2} \sum_\mu \Lambda_\mu^S \langle \phi_0(r_1) | r_{12}^{-1} | \Phi_\mu^S(r_1, r_2) \rangle , \qquad (5.52)$$

where $\mathcal{E}_\mu$ is the energy of the μ-th AIS.

Now consider the radial equation which is needed in some calculations. We shall represent the function $F_0^S(r)$ in the form of (3.65). Then substituting (3.65) and (5.14) into (5.51) and (5.52) we shall obtain

$$\left[\frac{d^2}{dr^2} - \frac{L(L+1)}{r^2} + \frac{2Z}{r} + k_0^2 - 2\langle\phi_0|r_{12}^{-1}|\phi_0\rangle\right] f_L^S(r)$$

$$+ 2(-1)^S \langle\phi_0|r_{12}^{-1}|f_L^S\rangle\phi_0 = \frac{2i^{-L}k_0}{(2\pi[L])^{1/2}} \sum_\mu \Lambda_\mu^S \chi_\mu^{LS}(r) , \tag{5.53}$$

$$\Lambda_\mu^S(E - \mathcal{E}_\mu) = (4\pi)^{1/2}\frac{i^L}{k_0}\int dr_2 \chi_\mu^{*LS}(r_2) f_L^S(r_2) , \tag{5.54}$$

where

$$\chi_\mu^{LS}(r_2) = (4\pi[L])^{1/2} \sum_{n_1 l_1 n_2 l_2} (-1)^{l_2 - l_1} \begin{pmatrix} L\ l_1\ l_2 \\ 0\ 0\ 0 \end{pmatrix} \beta_\mu(n_1 l_1 n_2 l_2)$$

$$\times r_2 \int r_1^2 dr_1 \phi_0(r_1)\frac{\gamma_{l_1}}{[l_1]} R_{n_1 l_1}(r_1) R_{n_2 l_2}(r_2) + (1 \rightleftarrows 2) . \tag{5.55}$$

Here $R_{nl}(r)$ is the radial part of the hydrogen-like functions $\phi_{nlm}(r)$; $[l] = 2l+1$; ϕ_0 is the wave function of the ground state He$^+$ ion. The presence of the slowly decreasing term $2Z/r$, which is due to the Coulomb interaction in (5.53) results in the following asymptotics for $f_L^S(r)$:

$$f_L^S(r) \sim \sin\left(k_0 r - \frac{\pi L}{2} + \frac{1}{k_0}\ln 2k_0 r\right)$$

$$+ \tilde{f}_L^S k_0 \exp\left[i\left(k_0 r + \frac{1}{k_0}\ln 2k_0 r\right)\right] . \tag{5.56}$$

Since the nonlocal exchange potentials in (5.53) can be replaced within a narrow energy range by a local potential [5.10], the solution of (5.53) with asymptotics in the form of (5.56) may also be found by the Green's function method

$$f_L^S(r_2) = u_L^S(r_2) + \int dr_2' G_L^S(r_2, r_2')\frac{2i^{-L}k_0}{(4\pi[L])^{1/2}} \sum_\mu \Lambda_\mu^S \Phi_\mu^{LS}(r_2') , \tag{5.57}$$

where

$$G_L^S(r, r') = \begin{cases} -k_0^{-1}\left(u_L^S(r)v_L^S(r') + iu_L^S(r')u_L^S(r)\right) , & r' < r \\ -k_0^{-1}\left(u_L^S(r')v_L^S(r) + iu_L^S(r')u_L^S(r)\right) , & r' > r \end{cases} \tag{5.58}$$

and u_L^S and v_L^S are solutions of (5.53) without the right-hand side with the following asymptotic behavior at infinity:

$$u_L^S(r) \sim \sin\left(k_0 r - \frac{\pi L}{2} + \frac{1}{k_0}\ln 2k_0 r + \delta_L^S\right) ;$$

$$v_L^S(r) \sim \cos\left(k_0 r - \frac{\pi L}{2} + \frac{1}{k_0}\ln 2k_0 r + \delta_L^S\right) . \tag{5.59}$$

Here δ_L^S is a potential scattering phase.

Substituting (5.57) into (5.54) and neglecting the nondiagonal terms like $\int dr_2 dr_2' \chi_\mu^{*LS}(r_2) G_L^S(r_2, r_2') \chi_\nu^{LS}(r_2)$ we obtain

$$\Lambda_\mu^S = (4\pi[L])^{1/2} i^L k_0^{-1} \int dr_2 \chi_\mu^{*LS}(r_2) u_L^S(r_2)$$

$$\times \left[E - \mathcal{E}_\mu - 2 \iint dr_2' dr'' \chi_\mu^{*LS}(r_2') \mathrm{Re}\, G_L^S(r_2', r_2'') \chi_\mu^{LS}(r_2'') \right.$$

$$\left. + \frac{2i}{k_2} \left| \int dr_2 \chi_\mu^{LS}(r_2) u_L^S(r_2) \right|^2 \right]^{-1} . \tag{5.60}$$

Substituting (5.57) into (5.60), and comparing the asymptotics with (5.56), we arrive at the explicit form of the partial amplitude

$$\tilde{f}_L^S = \frac{e^{2i\delta_L^S} - 1}{2ik_0} - (4\pi[L])^{1/2} e^{2i\delta_L^S} \sum_\mu \frac{\Gamma_\mu}{E - \mathcal{E}_\mu - \Delta_\mu + i\Gamma_\mu/2}, \tag{5.61}$$

where

$$\Gamma_\mu = \frac{4}{k_0} \left| \int_0^\infty dr_2 \chi_\mu^{LS}(r_2) u_L^S(r_2) \right|^2 ; \tag{5.62}$$

$$\Delta_\mu = 2 \iint dr_2 dr_2' \chi_\mu^{*LS}(r_2) \mathrm{Re}\, G_L^S(r_2, r_2') \chi_\mu^{LS}(r_2') . \tag{5.63}$$

Here $\delta_L^S = \eta_L + \sigma_L^S$; S is the total spin of the μ-th resonance; η_L is the Coulomb phase shift, whereas

$$\eta_L + \arg \Gamma(L + 1 - i/k_0) , \tag{5.64}$$

and the phase shift σ_L^S which is caused by the short-range potential is found by integration of (5.53) without the right-hand side under the boundary conditions

$$f_L^S(0) = 0 ; \quad f_L^S(r \to \infty) \sim \frac{F_L(k_0 r)}{k_0} + \tilde{f}_L^S [G_L(k_0 r) + i F_L(k_0 r)] . \tag{5.65}$$

Here $F_L(k_0 r)$ and $G_L(k_0 r)$ are the regular and nonregular Coulomb functions, respectively; the summation in (5.61) runs over those resonances which are allowed by the total spin and parity conservation laws. Thus, the first term in (5.61) describes the potential scattering; while the second one describes the resonance scattering.

The total scattering amplitude is given by (2.19); while the differential cross section is given by (2.27, 3.58). The direct use of (2.27) is, however, inconvenient since the partial wave expansion converges slowly due to the Coulomb field. If one confines himself to a single term in the sum over the resonance contributions in the expression for the amplitude (5.61), then the cross section may be represented as [5.10]

$$\frac{d\sigma^S}{d\Omega} = \frac{Z^2}{4k_0^4 \sin^4(\theta/2)} |1 + N^S|^2 , \tag{5.66}$$

where

$$N^S = \frac{2k_0}{Z} \sin^2 \frac{\theta}{2} \exp\left(-\frac{iZ}{k_0} \ln \sin^2 \frac{\theta}{2}\right) \sum_L (2L+1)$$

$$\times e^{2i(\eta_L - \eta_0)} e^{i(\sigma_L^S + \delta_L^r)} \sin\left(\sigma_L^S + \delta_L^r\right) P_l(\cos\theta) ;$$

$$\delta_L^r = \arctan \frac{\Gamma_\mu/2}{E - \mathcal{E}_\mu - \Delta_\mu} ;$$

δ_L^r is a resonance scattering phase which corresponds to that μ-th AIS which has a total angular momentum L, total spin S, parity $(-1)^L$ and lies near the given energy $E = k_0^2/2 + \mathcal{E}_0$. Note that (5.66) may be applied within the limited energy range ($E \sim \mathcal{E}_\mu$) which contains a single resonance.

If the calculations are carried out when L is given, then all the states with $L = l_1 + l_2$ must be considered. Thus, at $L = 0$, $l_1 + l_2 = 0$, one takes into account the states which satisfy these conditions, i.e., $2s^2$, $2s3s$, $2p^2$, $2p3p$ states, etc. When $n_1 = 2$, all these states are converging to $n = 2$. Since two-electron states may have a zero spin (singlets) or $S = 1$ (triplets) with a positive or negative parity, for $L = 0$ we have a ^{1,3}S-state, for $L = 1$, ^{1,3}P^0-state, and so on (the upper left superscript is $2S + 1$). We shall denote the states with odd parity by a superscript "0", otherwise, we mean that the state has even parity.

Let us now discuss the obtained results starting with the ^{1}P^0-resonances. Let us confine ourselves to the contribution of those AIS which have the total orbital angular momentum $L = 1$, total spin S, parity (-1), i.e., the 2s2p, 2s3p, 2p3s, 2p3d, 2s4d, 2p4s, 2p4d, 2s5p, 2p5s and 2p5d states. With this basis, the quantum numbers $(n_1 l_1, n_2 l_2)$ span the indicated values, while μ and ν cover the values from 1 to 10. This basis allows one to calculate the energies $\mathcal{E}_\mu$ and coefficients β_μ by solving the system (5.16) which, in turn, enables us to determine Γ_μ and Δ_μ from (5.62) and (5.63), respectively.

The estimations show that the corrections to the energies of AIS are small ($10^{-3} - 10^{-6}$ eV). Therefore, we shall neglect these corrections in further numerical calculations.

Table 5.1 presents the calculated and experimentally determined energies of the lowest AIS [5.5]. The relative positions of these states with respect to other levels are shown in Fig. 5.1. The resonance positions are presented on a scale where the energy of the ground state of an He atom is taken as the zero point

$$E[\text{eV}] = E_\text{i} + \mathcal{E}$$

where E_i is the double ionization potential for an He atom ($E_\text{i} = 79.0058$ eV), $\mathcal{E}$ is the energy on the scale where the energy for the doubly ionized He atom is taken to be zero.

Table 5.1 presents the data in the AIS classification scheme proposed in [5.15] using three quantities N, n, α. Here N denotes the He$^+$ level below which the AIS lies; n is the index of the AIS within a given series; α labels the series of AIS with a given term and assumes the values "a", "b", "c", etc. This classification scheme was devised to better sort out the states in view of the strong configurational mixing within the wave functions of He AIS. In this

Table 5.1. Energies [eV] of the $(2, n\alpha)^{2S+1}L$ AIS of the He atom

Classi-fication	Theory				Experiment	
	DM [5.5]	DM [5.3]	CI [5.12]	CCM [5.13]	[5.11]	[5.14]
(2,2a) ^{1}S	57.91	57.93	57.92	57.86	57.82	57.82
(2,2b)	62.27	62.31	62.78	62.81	62.15	62.06
(2,3a)	63.00	63.02	63.05	63.01	62.95	62.94
(2,4a)	64.20	64.21	64.23	64.22	64.22	64.18
(2,3a) ^{3}S	62.62	62.63	62.63	62.62	—	—
(2,3b)	63.79	63.79	63.82	63.82	—	—
(2,4a)	64.08	64.09	64.08	64.08	—	—
(2,2a) ^{1}P^0	60.27	60.35	60.33	60.27	60.10	60.13[a]
(2,3b)	62.77	62.79	62.78	62.77	—	62.76[a]
(2,3a)	63.69	63.71	63.70	63.69	63.65	63.66[a]
(2,4b)	64.14	64.14	64.14	64.13	—	64.14[a]
(2,2a) ^{1}P^0	58.37	58.41	58.39	58.36	58.34	58.30
(2,3a)	63.13	63.17	63.16	63.14	63.08	63.07
(2,3b)	63.26	63.29	63.28	63.28	—	—
(2,4a)	64.25	64.27	64.27	64.26	64.22	64.23
(2,2a) ^{1}D	60.03	60.06	60.09	—	60.00	59.89
(2,3a)	63.56	63.60	63.59	—	—	63.50
(2,3b)	63.87	63.90	63.90	—	—	—
(2,4a)	64.42	64.48	64.48	—	—	64.39
(2,3a) ^{3}D	63.14	63.15	63.15	—	—	—
(2,3b)	63.76	63.80	63.79	—	—	—
(2,4a)	64.27	64.28	64.28	—	—	—

[a] Photoabsorption experiments [5.11, 12]

case no certain configuration can be attributed to a specific AIS since the AIS with a given term are grouped into series differing in the type of the initial configuration mixing, in quantum defects and in the reduced widths. The latter are, in accordance with [5.16], quasiconstants for all terms within a given AIS series but differ for AIS series with different terms.

The results which are obtained by the DM agree well with the data calculated by the more complex CCM [5.13]. In [5.5] exchange is taken into account as well as the Coulomb field distortion due to the screening of the nucleus by an atomic electron. Inclusion of the exchange significantly affects the u_L^S-function and, hence, the potential phase shifts of the scattering. However, inclusion of the exchange does not influence, in practice, the resonance widths values which result from the fact that the difference in u_L^S in (5.62) is smoothed out by the integral.

Indeed, we obtain, e.g., the width of the 2s2p ^{1}P^0 resonance ($\Gamma = 3.54 \times 10^{-2}$ eV), which coincides with the results of [5.2] to within an accuracy of 1 %. The phases σ_L^S caused by the short-range potential are found by means of numerical integration of the homogeneous equation (5.53).

Fig. 5.1

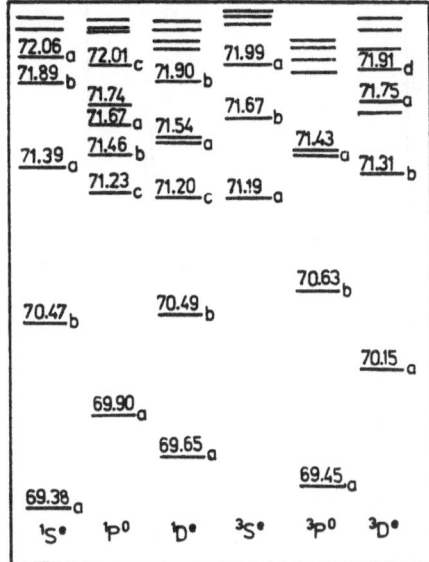

Fig. 5.2

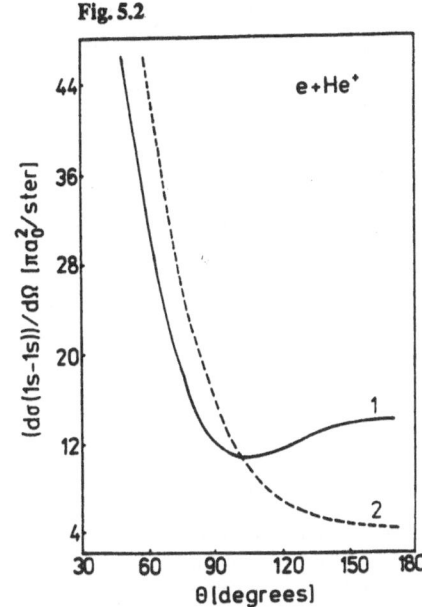

Fig. 5.1. Autoionizing levels of He converging to the $n = 3$ threshold of He$^+$. Indices a, b, c indicate the series to which the given level belongs. Numbers indicate the energies in eV

Fig. 5.2. Angular dependence of the differential cross section of e + He$^+$ elastic scattering at an incident electron energy of 3 eV. *Curve 1* – calculation according to (5.66), *curve 2* – calculation according to the Rutherford formula (without accounting for short-range interaction)

The phase shifts of the potential scattering of an electron by a helium ion have been computed by *Drukarev* [5.17] (for more complete calculations see Ref. 5.5). These phase shifts as well as the resonance parameters allow one to determine the differential cross sections according to (5.66).

As can be seen from Fig. 5.2, the influence of the short-range potential is quite significant especially at large angles. The energy dependences of differential cross sections at $\theta = 45°$ and $\theta = 170°$ are shown in Fig. 5.3.

Let us point out what the contribution of lower resonance to the differential cross section is. A more detailed study of the behaviour of the structure in the cross section shows that at small angles the S-resonances form dips (i.e., destructive interference between the potential and resonance scattering occurs). It is seen, for example, that the $2s^2$ ^1S autoionizing state (2, 2a) according to the classification suggested in Ref. [5.5] with an energy of $E = 57.91$ eV arises as a resonance dip at an incident electron kinetic energy equal to $57.91 - I = 33.4$ eV, and corresponds to $k = 1.57$ au (note that $I = 24.5$ eV is a single ionization potential for a helium atom). At the same time the contribution of the P-resonance has a symmetric form, i.e. forms a peak and a dip, while the D-resonance results in the formation of a peak. It has been seen that near $\theta = 170°$ the height of the P-resonance decreases while that of the S-resonance becomes symmetric. These features are shown in Fig. 5.3a,b.

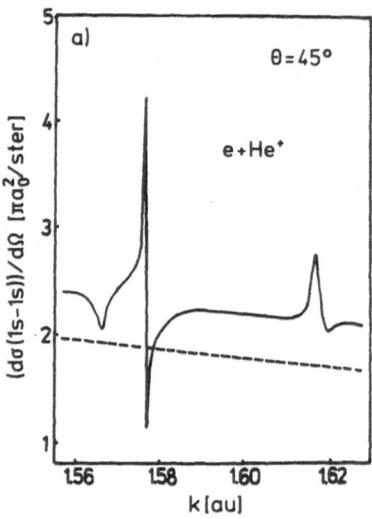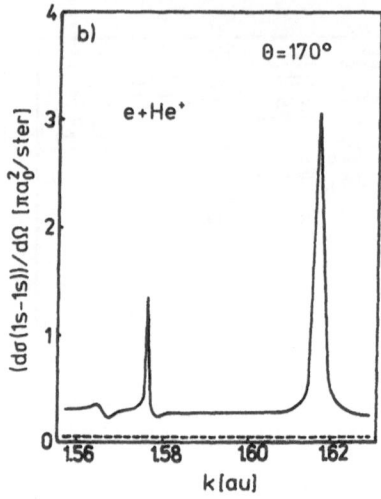

Fig. 5.3a,b. Energy dependence of the differential cross section of e + He$^+$ elastic scattering in the region of the lower ^{1}S (at $k \approx 1.57$ au), $^3P^0$ (at $k \approx 1.58$ au), ^{1}D (at $k \approx 1.62$ au) resonances at two different scattering angles θ [5.5]

In inelastic scattering, i.e., electron impact excitation of He$^+$ levels is of particular interest. Section 7.4 deals with the examination of resonances in inelastic scattering.

The AIS considered here generate resonances not only in the process of electron scattering by a helium atom but also in photoionization and electron impact ionization of a He atom, as well as in dielectron recombination of the He$^+$ ion. The study of these processes goes beyond the scope of the present treatment. For detailed information on this subject we refer the reader to the original papers [5.2, 8, 19].

Let us note that the difference between curves 1 and 2 on Fig. 5.2 is in fact due to the complicated structure of the target. Exactly similar difference in electron-proton scattering at ultra-high energies have suggested a dramatic idea to Feynman that proton is consisting of partons (quarks in fact).

6. Multichannel Resonance Scattering of Electrons by Complex Atoms

If the energy of the incident electron is such that two or more channels are open, the AIS which pass through a region of multichannel decay can be generated. In that case, resonances may arise both in elastic and inelastic channels. In this chapter such multichannel resonances are discussed. Resonances in the electron scattering by complex atoms and ions are emphasized.

6.1 Complex-Atom Wave Functions

The description of electron scattering by hydrogen atoms considered above can be extended to electron scattering by complex atoms. However, unlike the case for hydrogen-like atoms, there exist no exact analytical expressions for the wave functions of complex target atoms. Nevertheless, quantum mechanics provides quite effective approximate numerical methods to find the wave functions. These methods yield, at least for not very heavy atoms, quite satisfactory results. Out of the perturbation method, variational method and the Hartree-Fock method the latter one gives, in some cases, the best results [6.1]. This method provides excellent results for light atoms, however, in general the larger the number of electrons, the higher the deviation of the Hartree-Fock results from experiment.

The main assumptions in the Hartree-Fock approach are i) the factorization of the wave function and ii) that the potential is central.

It follows from the second assumption that all electrons in the same shell are equivalent. Note that this statement is valid both for filled and half-filled shells. Thus, the application of the Hartree-Fock method to calculate atoms of other types requires an additional modifiaction. The first assumption means that part of the inter-electron interaction is not taken into account. This part is usually called the electron correlation and obviously, is equal to the difference between the total inter-electron interaction $V^{\mathrm{tot}} = \sum_{i \neq f} 1/ \mid r_i - r_f \mid$ and the inter-electron interaction calculated by the Hartree-Fock method

$$V^{\mathrm{HF}} = \sum_{i \neq f} \int \frac{dr_i \mid \phi_i(r_i) \mid^2}{\mid r_i - r_f \mid} + V_{\mathrm{e}} , \qquad (6.1)$$

which takes into account only the averaged mutual electron interaction. Here V_{e} is the nonlocal exchange part of the Hartree-Fock potential.

A variety of methods which allow one to take into account at least part of the inter-electron correlation have been developed. We have already mentioned the CI method, the MultiConfiguration Hartree-Fock method (MCHF), the Random Phase Approximation with Exchange (RPAE), the Hyperspherical Coordinate Method (HSC) and others.

In the first method, which we considered earlier in an example of a two-electron system (Chap. 5), one looks for the wave function of the system not as the determinant corresponding to the considered configuration but rather as a linear combination of those determinants which correspond to all possible atomic configurations:

$$\Psi(\gamma LS) = \sum_{i=1}^{N} \beta_i \Psi(\gamma_i LS) \equiv \beta \Psi , \tag{6.2}$$

where γ_i defines the configuration and $\Psi(\gamma_i LS)$ are considered to be of the Hartree-Fock type. The energy is found from

$$E = \langle \Psi(\gamma LS) \mid H \mid \Psi(\gamma LS) \rangle \equiv \beta^+ H \beta ; \tag{6.3}$$

where

$$H_{ij} = \langle \Psi(\gamma_i LS) \mid H \mid \Psi(\gamma_j LS) \rangle , \quad \beta^* \beta = 1 .$$

If one chooses the functions $\Psi(\gamma_i LS)$ to satisfy the single-configuration equations, then the coefficients β_i are found from

$$H_{ij} \beta_j = E \beta_i ,$$

or in more detail

$$\sum_{i}^{M} \left(\langle \Psi(\gamma_j LS) \mid H \mid \Psi(\gamma_i LS) \rangle - \delta_{ij} E \right) \beta_i = 0 , \quad j = 1, \dots , M . \tag{6.4}$$

The latter allows one to obtain more accurate energy values and wave functions. In practice, of course, the sum (6.4) is restricted only to several terms which correspond to the neighbouring configurations.

Within the MCHF method [6.2] both the coefficients β_i and the wave functions $\Psi(\gamma_i LS)$ are determined simultaneously from (6.4) assuming $\Psi(\gamma_i LS)$ to be known (e.g., according to Hartree-Fock) and β_i to be equal to δ_{i1}. Then the new β and Ψ are sought, and the procedure is repeated.

Let us apply this method to a helium atom. It is quite sufficient to retain only two terms

$$\Psi(\gamma LS) = \beta_1 \Psi(1s^2 \, {}^1S) + \beta_2 \Psi(1s2s \, {}^1S) . \tag{6.5}$$

Then for radial functions

$$\hat{D}P_{1s}(r) = \frac{2}{q(1s)r} \left[\beta_1^2 Y_0(1s2s; r) + 2\beta_2^2 Y_0(1s1s; r) \right] P_{1s}(r)$$

$$+ \frac{2}{q(2s)r} \beta_1^2 Y_0(1s2s; r) P_{2s}(r) + \frac{2(2)^{1/2}}{q(1s)r} \beta_1 \beta_2 \left[2 Y_0(1s2s; r) P_{1s}(r) \right.$$

$$\left. + \mathcal{E}_{1s,2s} P_{2s}(r) + Y_0(1s1s; r) P_{2s}(r) - \frac{r}{2} \hat{D} P_{2s}(r) \right]$$

$$+ \mathcal{E}_{1s,1s} P_{1s}(r) \; ; \qquad\qquad (6.6)$$

$$\hat{D}P_{2s}(r) = \frac{2}{r} \left(Y_0(1s1s; r) P_{2s}(r) + Y_0(1s2s; r) P_{1s}(r) \right.$$

$$+ \frac{2(2)^{1/2}}{r} \frac{\beta_2}{\beta_1} \left(Y_0(1s1s; r) P_{1s}(r) - \frac{r}{2} \hat{D} P_{1s}(r) \right)$$

$$+ \mathcal{E}_{2s,2s} P_{2s}(r) + \mathcal{E}_{2s,1s} P_{1s}(r) \; , \qquad\qquad (6.7)$$

where, as usual $\hat{D} = d^2/dr^2 - l(l+1)/r^2 + 2Z/r$. The other notations are borrowed from Hartree's monograph [6.1].

This system must be solved jointly with the secular equation for β. The method provides quite accurate values for energies and wave functions.

The results of the numerical solution of the system (6.6, 7) for the ground He(^{1}S state are presented in [6.2], where $E_{tot} = -5.72$ Ry (77.8 eV) is obtained. With eleven configurations $1s^2$, $2s^2$, $3s^2$, $4s^2$, $2p^2$, $3p^2$, $4p^2$, $3d^2$, $4d^2$, $4f^2$, $5g^2$ taken into account we have $E_{tot} = -5.81$ Ry (79.04 eV), which coincides quite well with the experimental spectroscopic data. From a comparison of these results with those presented in Sect. 3.3 we can get an idea about the accuracy of the different methods.

The RPAE method takes into account a large part of inter-electron correlation interaction. This method, as shown in [6.3], is equivalent to the time-dependent Hartree-Fock approximation. Within the language of perturbation theory, the third method summarizes an infinite number of Feynman diagrams of the same type which describe the dynamic polarization of the atom. Numerous calculations within the RPAE method have been performed by *Amusia* and colleagues [6.3].

Each of the methods considered above has its own advantages and short-comings. A common feature is that they do not allow one to take into account the inter-electron interaction in a consistent way since each of them is based on the single-particle approximation. This leads to the slow convergence as well as to difficulties in the classification of the states under consideration. These difficulties are avoided within the HSC (hyper-spherical coordinates) method due to the introduction of collective coordinates [6.4]. The HSC method has been applied in atomic physics to calculate the energies of He and He$^-$ [6.5–7]. It was generalized to a system of two electrons outside a filled shell as well as to three-electron systems [6.8–11].

Let us briefly review the HSC method. In studying the AIS of the He atom [6.12] it has been established that the widths of those states which converge to the $n = 2$ threshold and belong to different series differ approximately by a factor of 100. To explain the existence of these series of AIS, it had been suggested in [6.12] to describe them by the wave functions

$$| 2n\pm\rangle = \frac{1}{2^{1/2}} \left(| 2snp \, {}^1P^0\rangle \pm | 2pns \, {}^1P^0\rangle \right) , \qquad (6.8)$$

where $|2snp^1P^0\rangle$ and $|2pns^1P^0\rangle$ are the single-configuration functions. The probability of both electrons being located near the nucleus is much greater for $|2n+\rangle$-states than for $|2n-\rangle$-states. The result is that the probability of a three-body collision (between a nucleus and both electrons) and of the transfer of energy sufficient for one electron to leave the atom is also much greater for "+"-states. Hence, the probability of autoionization and, thus, the width of the states are greater for "+"-series than those for "−"series.

The strong configuration mixing in AIS established here specifies that the properties of these states are defined, to a considerable extent, by the correlations in the motion of the excited electrons in the atom. The role of the correlation increases significantly in the autodetaching states of negative ions which arise in the process of electron scattering by neutral atoms.

Consider, for example, an application of the HSC method to the problem mentioned above. Let us introduce the collective variables for the single-electron coordinates r_1 and r_2

$$R = \left(r_1^2 + r_2^2 \right)^{1/2} , \quad \alpha = \arctan \frac{r_1}{r_2} ,$$
$$\hat{r}_1 = \frac{r_1}{r_1} , \quad \hat{r}_2 = \frac{r_2}{r_2} . \qquad (6.9)$$

Now it is possible to state that instead of the single-electron coordinates which describe the independent motion of two electrons in three-dimensional space, the HSC method utilizes the motion of collective coordinates which describe the motion of a single point representing an electron pair in a six-dimensional space.

If one introduces the function $\Psi = R^{5/2} \sin \alpha \cos \alpha \mathcal{Y}$ for the $\mathcal{Y}$-function of a two-electron system, then the Schrödinger equation in the new variables becomes

$$\left[\frac{1}{2} \left(\frac{\partial^2}{\partial R^2} + \frac{\hat{\Lambda}^2}{R^2} \right) + V(R, \alpha, \hat{r}_1, \hat{r}_2) - E \right] \Psi(R, \alpha, \hat{r}_1, \hat{r}_2) = 0 , \qquad (6.10)$$

where V is the potential energy of interaction of an electron with the nucleus and with another electron, $\hat{\Lambda}^2$ is the operator of the angular momentum squared in the six-dimensional space with $\lambda(\lambda + 4) + 15/4$ eigenvalues, $\lambda = l_1 + l_2 + 2n$, $n = 0, 1, 2, \ldots$.

Let us solve (6.10) within the so-called adiabatic approximation [6.5–7]. In order to do this, we shall introduce for each fixed R a system of orthogonal functions $\Phi_\mu(R; \omega)$, $\omega \equiv (\alpha, \hat{r}_1, \hat{r}_2)$ being the solution of the following equations for eigenfunctions and eigenvalues

$$\hat{U}\Phi_\mu(R, \omega) = U_\mu(R)\Phi_\mu(R, \omega) , \quad \hat{U} = \hat{\Lambda}^2/2R^2 - V . \qquad (6.11)$$

Here $U_\mu(R)$ are the eigenvalues of the operator U. Expanding Ψ into a series

$$\Psi = \sum_\mu F_\mu(R)\Phi_\mu(R;\omega) \tag{6.12}$$

and using (6.10), we obtain a system of radial equations for the functions F_μ

$$\left[-\frac{d^2}{dR^2} + U_\mu(R) + 2E\right] F_\mu(R) = \sum_\nu \hat{P}_{\mu\nu} F_\nu(R) , \tag{6.13}$$

where

$$\hat{P}_{\mu\nu} = 2\int d\omega \Phi_\mu^*(R;\omega)\frac{\partial}{\partial R}\Phi_\nu(R;\omega)\frac{\partial}{\partial R}$$

$$+ 2\int d\omega \Phi_\mu^*(R;\omega)\frac{\partial^2}{\partial R^2}\Phi_\nu(R;\omega) . \tag{6.14}$$

The nondiagonal elements of the P-matrix are neglected in the adiabatical approximation, i.e., the coupling between different channels μ in (6.13) is eliminated.

Figure 6.1 shows potentials $U_\mu(R)$ for the helium atom. As can be seen, some of the curves reveal a distinct well. These potentials may result in the formation of a bound state in a given channel labeled by the index μ. If the potentials $U_\mu(R)$ are known one can solve (6.13) for eigenfunctions and eigenvalues E. The negative eigenvalues define the positions of the AIS. Every potential U_μ generates a series of AIS, e.g., the U_2 potential, according to the classification presented in [6.12], generates the $(2, na)^1 S(l_1 = l_2 = 0)$ states of a He atom converging to the $n = 2$ threshold of the He$^+$ ion. Thus, it is seen that the HSC method allows one not only to obtain the parameters of AIS but to classify them into series.

Note that the consideration of relativistic corrections is important at least for heavy atoms. These corrections are usually proportional to a factor $\alpha^2(\alpha = e^2/\hbar c \approx 1/137)$. For example, these corrections are necessary for the mercury atom or heavier species. The same holds true for the calculation of the inner shells of even light atoms. The relativistic corrections may be calculated to excellent accuracy if one proceeds not from the Schrödinger equation but from the Dirac equation when obtaining the Hartree-Fock equation.

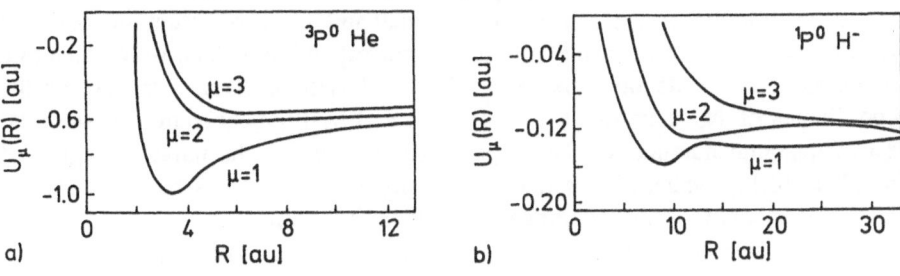

Fig. 6.1. Potentials $U_\mu(R)$ for the He atom (a) and H$^-$ ion (b)

6.2 Initial System of Equations

Above, we have obtained the equations which describe the processes of resonance scattering of electrons by the simplest atoms and ions in the single-channel approximation. Such an approach can describe only the resonances in an elastic cross section. Resonances in an excitation cross section can be described only within multichannel theory. Thus, we shall generalize the obtained results in two directions: we shall analyze the electron scattering by atoms of arbitrary complexity and also write the system of equations for an arbitrary number of channels.

The Hamiltonian H which describes the electron scattering by an N-electron atomic target with nuclear charge Z is

$$H = H_N - \frac{1}{2}\nabla^2_{N+1} - \frac{Z}{r_{N+1}} + \sum_{i=1}^{N} \frac{1}{r_{i,N+1}} , \tag{6.15}$$

where H_N is the Hamiltonian of the target

$$H_N = \sum_{i=1}^{N} \left(-\frac{1}{2}\nabla^2_i - \frac{Z}{r_i}\right) + \sum_{i<j} \frac{1}{r_{ij}} . \tag{6.16}$$

The scattering problem is thus the solution of the Schrödinger equation

$$(H - E)\Psi(X, \boldsymbol{x}_{N+1}) = 0 \tag{6.17}$$

with corresponding boundary conditions. The function $\Psi(X, \boldsymbol{x}_{N+1})$, which is usually called the collision wave function, is a completely antisymmetrized wave function of the system "incident electron + target"; $X \equiv (\boldsymbol{x}_1, \ldots, \boldsymbol{x}_N)$, $\boldsymbol{x}_i \equiv (\boldsymbol{r}_i, \sigma_i)$ defines a family of spatial $\boldsymbol{r}_i$ and spin σ_i coordinates of the i-th electron.

Consider now the structure of the function $\Psi(X, \boldsymbol{x}_{N+1})$. We shall start with the nonsymmetrized wave function of the N-electron target

$$\Psi_u(q\gamma_{\mathrm{T}}X) = \left[\prod_\lambda \left(q_\lambda \mid n_\lambda l_\lambda^{N_\lambda} \alpha_\lambda S_\lambda L_\lambda \right) \right]^{\gamma_{\mathrm{T}}} , \tag{6.18}$$

where $(q_\lambda | n_\lambda l_\lambda^{N_\lambda} \alpha_\lambda S_\lambda L_\lambda)$ is a completely antisymmetrized wave function of the electron subshell λ; n_λ, l_λ and N_λ are principal and orbital quantum numbers and the number of electrons in the subshell λ, correspondingly. The set of quantum numbers $\alpha_\lambda S_\lambda L_\lambda$ denotes the term of the subshell; q denotes the set of the coordinates of N_λ electrons ($q_\lambda \equiv \{r_i, \sigma_i\}$) and γ_{T} does the same for the total set of quantum numbers of the target. Hereafter we use the notation suggested by *Fano* [6.13] which allows one to simplify the expressions for the complex matrix elements to a quite compact and convenient form.

The nonsymmetrized collision wave function can be written as an expansion over the complete set of the target wave functions:

$$\Psi_u(qX\boldsymbol{x}_{N+1}) = \sum_{\gamma_T} \Psi_u(q\gamma_T X) \tilde{F}_{\gamma_T}(X_{N+1}) . \qquad (6.19)$$

Here the summation in (6.19) also means integration over the continuous spectrum of the target states.

If we represent $\tilde{F}_{\gamma_T}(\boldsymbol{x}_{N+1})$ in the form

$$\tilde{F}_{\gamma_T}(X_{N+1}) = \sum_{l_T m_T m_s} \frac{f_{\gamma_T}^{l_T m_T m_s}(r_{N+1})}{r_{N+1}} Y_{l_T m_T}(r_{N+1}) \chi_{m_s}^{1/2}(\sigma_{N+1}) \qquad (6.20)$$

where $\chi_{m_s}^{1/2}(\sigma_{N+1})$ is a spin function of the $(N+1)$-th electron, then the nonsymmetrized collision wave function may be written as:

$$\Psi_u(qX\boldsymbol{x}_{N+1}) = \sum_{\Gamma} \Psi_u(q\Gamma X \hat{r}_{N+1}\sigma_{N+1}) \frac{\tilde{F}_{\Gamma}(r_{N+1})}{r_{N+1}} , \qquad (6.21)$$

where:

$$\Psi_u(q\Gamma X \hat{r}_{N+1}\sigma_{N+1}) = \left[\Psi_u(q\gamma_T X) \times (N+1 \mid k_T l_T \tfrac{1}{2}) \right]^{\Gamma} , \qquad (6.22)$$

$$\tilde{F}_{\Gamma}(r_{N+1}) = \sum_{m_s m_T} (L_T M_T l_T m_T \mid LM)(S_T M_{S_T} \tfrac{1}{2} m_s \mid SM_S)$$
$$\times f_{\gamma_T}^{l_T m_T m_s}(r_{N+1}) . \qquad (6.23)$$

In (6.20–23), $\Gamma = (\gamma_T l_T LM S M_S)$ denotes the complete set of quantum numbers of the system of $(N+1)$ electrons, $L_T M_T S_T M_{S_T}$ are the total angular momentum and spin of the target and their projections, $l_T m_T m_s$ are the angular momentum and spin of the incident electron with respect to the target, $\times$ denotes the vector cross product of the wave function $\Psi_u(q\gamma_T X)$ with the spin-angular function $(N+1 \mid k_T l_T \tfrac{1}{2})$ of the $(N+1)$-th electron.

Consider now the boundary conditions which the radial functions $\tilde{F}_{\Gamma}(r)$ must obey. In the asymptotic region, $\tilde{F}_{\Gamma}(r)$ is a superposition of converging and diverging spherical waves:

$$\tilde{F}_{\Gamma}(r \to \infty) = A_{\Gamma}e^{-i\theta_{\Gamma}} + B_{\Gamma}e^{i\theta_{\Gamma}} , \qquad (6.24)$$

where:

$$\theta_{\Gamma} = k_{\Gamma}r - \frac{\pi l_{\Gamma}}{2} + \frac{Z-N}{k_{\Gamma}} \ln 2k_{\Gamma}r + \sigma_{l_{\Gamma}} ,$$

$$\sigma_{l_{\Gamma}} = \arg \Gamma \left(l_{\Gamma} + 1 - i\frac{Z-N}{k_{\Gamma}} \right) .$$

The coefficients A_{Γ} and B_{Γ} define the S-matrix of the scattering process:

$$B_{\Gamma} \equiv \sum_{\Gamma'} S_{\Gamma\Gamma'} A_{\Gamma'} . \qquad (6.25)$$

Introducing the new functions $F_{\Gamma\Gamma'}(r)$ by means of the relation

$$\tilde{F}(r) = \sum_{\Gamma'} F_{\Gamma\Gamma'}(r) \underset{r\simeq\infty}{\,} \sum_{\Gamma'} A_{\Gamma'} \left(\delta_{\Gamma\Gamma'} e^{-i\theta_\Gamma} - S_{\Gamma\Gamma'} e^{i\theta_\Gamma} \,\right) \tag{6.26}$$

we shall rewrite (6.21) as

$$\Psi_u(q\boldsymbol{X}\boldsymbol{x}_{N+1}) = \sum_{\Gamma\Gamma'} \Psi_u(q\Gamma X \hat{r}_{N+1}\sigma_{N+1}) \frac{F_{\Gamma\Gamma'}(r_{N+1})}{r_{N+1}} . \tag{6.27}$$

If the system "target + incident electron" has initially occupied the Γ' quantum state (i.e., Γ' is a reaction entrance channel), then the nonsymmetrized collision wave function is

$$\Psi_u\left(q\Gamma' \boldsymbol{X}\boldsymbol{x}_{N+1}\right) = \sum_{\Gamma} \Psi_u(q\Gamma X \hat{r}_{N+1}\sigma_{N+1}) \frac{F_{\Gamma\Gamma'}(r_{N+1})}{r_{N+1}} . \tag{6.28}$$

Let us now construct the completely antisymmetrized collision wave function. According to [6.14], we shall construct, first of all, a totally antisymmetrized wave function of the target:

$$\Psi(\gamma_{\mathrm{T}}\boldsymbol{X}) = D(N_\lambda)^{-1/2} \sum_q (-1)^{p_q} \Psi_u(q\gamma_{\mathrm{T}}\boldsymbol{X}) , \tag{6.29}$$

where $D(N_\lambda) = N!/\prod_\lambda N_\lambda!$ is the number of different distributions of N electrons over the subshells, p_q defines the parity q of the distribution. Carrying out the antisymmetrization of the incident electron with respect to all target electrons, we obtain finally the completely antisymmetrized collision wave function

$$\Psi\left(\Gamma'\boldsymbol{x}_1 \ldots \boldsymbol{x}_{N+1}\right)$$
$$= (N+1)^{-1/2} \sum_{p_q}^{N+1} (-1)^{N+1-p} \sum_{\Gamma} \Psi\left(\Gamma X \hat{r}_p \sigma_p\right) \frac{F_{\Gamma\Gamma'}(r_p)}{r_p} , \tag{6.30}$$

where the function

$$\Psi\left(\Gamma X \hat{r}_p \sigma_p\right)$$
$$= D(N_\lambda)^{-1/2} \sum_q (-1)^{p_q} \left[\left[\prod_\lambda (q_\lambda \mid n_\lambda l^{N_\lambda} \alpha_\lambda S_\lambda L_\lambda \} \right] (p \mid k_{\mathrm{T}} l_{\mathrm{T}} \tfrac{1}{2} \} \right]^{\Gamma}$$

contains not only the target wave function but also the spin-angular part of the scattered electron wave function. In the Close-Coupling Method (CCM) the sum (6.30) over Γ includes a finite number of open and closed channels.

The solutions of the Schrödinger equation are taken in the form of the wave function $\Psi(\gamma_{\mathrm{T}}\boldsymbol{X})$ of the target

$$\left(H_N - E_{\gamma_{\mathrm{T}}}\right) \Psi\left(\gamma_{\mathrm{T}}\boldsymbol{X}\right) = 0 . \tag{6.31}$$

Applying the variational principle to solve (6.31), one may obtain the system of

Hartree-Fock equations to find the wave functions of the ground and excited states of the target. However, the wave functions of the excited states thus constructed will not be orthogonal to that of the ground state since the wave functions of the ground and excited states are eigenfunctions of different Hamiltonians. This orthogonality may be obtained by using the "frozen core" approximation. Here the single-particle wave functions of the excited states are calculated in the field of the core whose wave functions are constant and are taken to be the same as for the ground state. Thus, the single-particle wave functions of both the ground and of the excited states will be automatically mutualy orthogonal since they are eigenfunctions of the same Hamiltonian. Hence, the total wave functions of the target will be mutually orthogonal too.

Using the variational principle to find the solutions of (6.17) where Ψ has the form given in (6.30), one may obtain the system of coupled integro-differential equations for the radial wave functions $F_{\Gamma\Gamma'}(r)$ of the continuum. This system may be solved by the well known methods. However, as was shown in [6.15], the expansion (6.30) does not define $F_{\Gamma\Gamma'}(r)$ uniquely which does often lead to a loss of accuracy in the numerical integration. To avoid the instability in the numerical solution, it is enough, as shown in [6.15], to require the radial wave functions $F_{\Gamma\Gamma'}(r)$ of the continuum to be orthogonal to the corresponding radial wave functions $P_{n_\lambda l_\lambda}(r)$ of the subshells of the target:

$$\int_0^\infty F_{\Gamma\Gamma'}(r)P_{n_\lambda l_\lambda}(r)dr = 0 \ . \tag{6.32}$$

The condition of (6.32) actually signifies that the incident electron cannot be virtually captured into one of the unfilled subshells which are taken into account in the expansion (6.30) of the target states.

However, once the orthogonality conditions are imposed, in order to obtain the same functions Ψ one should add the corresponding set of so-called correlation functions χ_μ to (6.30)

$$\chi_\mu \left(LS\pi x_1 \ \ldots \ x_{N+1} \right)$$
$$= D\left(N_\lambda^\mu\right)^{-1/2} \sum_q (-1)^{P_{q\mu}} \chi_u \left(q_\mu LS\pi x_1 \ \ldots \ x_{N+1} \right) \ , \tag{6.33}$$

where μ is an index labeling the correlation functions $\chi(LS\pi x_1 \ \ldots \ x_{N+1})$ (i.e., it assumes the values of the numbers of unfilled electron subshells which may capture the incident electron), $\chi_u(q_\mu LS\pi x_1 \ \ldots \ x_{N+1})$ is the nonsymmetrized wave function of the $(N+1)$-electron system which is constructed similarly to (6.18). Then the expansion (6.30) must be replaced by

$$\Psi_t \left(\Gamma_i x_1 \ \ldots \ x_{N+1} \right)$$
$$= \Psi \left(\Gamma_i x_1 \ \ldots \ x_{N+1} \right) + \sum_\mu C_\mu^{\Gamma_i} \chi_\mu \left(LS\pi x_1 \ \ldots \ x_{N+1} \right) \ . \tag{6.34}$$

Here $C_\mu^{\Gamma_i}$ are certain unknown expansion coefficients to be determined. The introduction of the correlation functions into (6.30) means that, in spite of the fact that (6.32) is valid, the expansion (6.34) takes into account the possibility of the virtual electron being captured into the unfilled subshells.

Using the variational principle, as done earlier, one can obtain [6.14, 16] the coupled system of the algebraic integro-differential equations for the radial functions $F_{\Gamma\Gamma'}$ and coefficients D_μ^Γ. Solving this system numerically one finds the radial functions and scattering matrices in the CCM.

The higher the number of included closed channels, the higher is the accuracy of the close-coupling calculations. However, the increase of the number of channels under consideration results in the growth of the number of equations for radial functions. It is obvious that this number is restricted by computational capabilities. Quite accurate results are obtained at lower energies only if the number of open channels is small too. One may restrict the system to a small number of virtual transitions. In most cases the contribution to the polarizability is given by the number of intermediate states, including the continuum. The application of the CCM to atoms of such type is complicated by the weak convergence of (6.34).

The accuracy of the calculations in this case can sometimes be improved without a great increase in the volume of computations by adding to expansion (6.34) a certain number of so-called pseudostates $\bar\phi_\gamma$. To some extent, the pseudostates are called upon to incorporate closed channels which are not incorporated in expansion (6.30). The functions $\bar\phi_\gamma$ are chosen in such a way that the family of the states $\Psi(\Gamma_i)$ and pseudostates $\bar\phi_\gamma$ which are taken into account leads to the experimentally observed value of the atomic polarizability. On the other hand, this approach has the disadvantage that the introduction of pseudostates sometimes gives rise to false resonances. As was noted in the previous chapters, an important role in the electron-atom scattering is played by the electron capture by a target into an AIS. The possibility of this process is taken into account in the CCM by the introduction of closed channels into the expansion in (6.34), which leads to a significant increase in the number of equations to be solved. In the DM, the only terms which are retained in expansion (6.30) are those which correspond to open channels. Closed channels and the possible capture of the incident electron by the target are dealt with by adding to the expansion the wave functions of the AIS which are multiconfiguration functions constructed by the CI method. These functions Φ_μ are constructed as a linear combination of single-configuration wave functions of an $(N+1)$-electron system.

The configuration mixing coefficients are found from condition (5.9). The basis of configurations from which the functions Φ_μ are constructed is chosen in such a way that the functions Φ_μ are orthogonal with respect to the open channel functions Ψ which are included in the sum over Γ in (6.30). The functions Φ_μ and the energies $\mathcal{E}_\mu$ thus represent the wave functions and energies of the AIS in the approximation in which the coupling between the open and closed channels is switched off. In the DM, this coupling is taken into account in the first-order perturbation theory; it gives rise to an energy shift of the AIS and to finite lifetimes for these states.

Adding the Φ_μ to (6.34) incorporates the closed channels only approximately, since the functions Φ_μ decrease exponentially with respect to all variables, while the functions $F_{\Gamma\Gamma'}$ in (6.3) may decrease far more slowly, e.g., in power-law fashion. For this reason, the polarization of the atom by the incident electron is incorporated in a slightly poorer way than in the CCM.

On the other hand, the parameters of the AIS may, on occasion, be found even more accurately in the DM, since it is possible, without any particular difficulty, to expand the basis of the wave functions from which Φ_μ are constructed. We will see an example of this situation in Chap. 7. Then the initial expansion of the collisional wave function within the DM approach has the form

$$\Psi_t(\Gamma_i x_1 \dots x_{N+1}) = \Psi(\Gamma_i x_1 \dots x_{N+1}) + \sum_\mu C_\mu^{\Gamma_i} \chi_\mu(LSx_1 \dots x_{N+1})$$

$$+ \sum_\nu \Lambda_\nu^{\Gamma_i} \Phi_\nu(LSx_1 \dots x_{N+1}) . \tag{6.35}$$

An approximate solution of (6.17) with Ψ being chosen as in (6.35) may be obtained by using the variational principle, namely, from the condition

$$\left(\Psi_t(\Gamma_k x_1 \dots x_{N+1}) \mid H - E \mid \Psi_t(\Gamma_l x_1 \dots x_{N+1}) \right) = 0 , \tag{6.36}$$

where Ψ_t is an approximate solution of (6.17) while the variation $\delta\Psi_t$ is obtained by varying $F_{\Gamma_k\Gamma_l}(r)$, $C_\mu^{\Gamma_k}$ and $\Lambda_\nu^{\Gamma_k}$ under the constraint (6.32). The variation of the radial wave function of the continuum is such that

$$\delta F_{kl} \underset{r \to \infty}{\sim} k_k^{-1/2} \delta K_{kl} \cos \theta_k , \tag{6.37}$$

where the reactance matrix K is determined from the open channel asymptotics

$$F_{kl}(r \to \infty) \sim k_k^{-1/2} [\delta_{kl} \sin \theta_k + (\cos \theta_k) K_{kl}] . \tag{6.38}$$

Upon transforming the matrix elements entering into $\langle \Psi_t \mid H - E \mid \Psi_t \rangle$ by the method described in [6.13, 14], we shall rewrite (6.36) in the following manner:

$$\delta \Bigg[\sum_{ij} \int F_{ik} L_{ij} F_{jl} dr + \sum_{j\mu} C_\mu^k \int V_{j\mu}^l F_{jl} dr$$

$$+ \sum_{i\nu} C_\nu^l \int V_{i\nu}^l F_{ik} dr + \sum_{\mu\nu} C_\mu^k C_\nu^l (A_{\mu\nu} - E\delta_{\mu\nu})$$

$$+ \sum_{j\mu} \Lambda_\mu^K \int U_{j\mu} F_{jl} dr + \sum_{i\nu} \Lambda_\nu^l \int U_{i\nu} F_{ik} dr + \sum_{\mu\nu} C_\mu^k \Lambda_\nu^k W'$$

$$+ \sum_{\nu\mu} \Lambda_\mu^K C_\nu^l W'_{\nu\mu} + \sum_{\mu\nu} \Lambda_\mu^K \Lambda_\nu^l (\mathcal{E}_\mu - E) \delta_{\mu\nu} - \tfrac{1}{2} K_{kl} \Bigg] = 0 , \tag{6.39}$$

where

$$L_{ij} = -\frac{1}{2} \left[\frac{d^2}{dr^2} - \frac{l_i(l_i + 1)}{r^2} + \frac{2Z}{r} + (E - \mathcal{E}_i) \right] \delta_{ij} + V_{ij} + W_{ij} .$$

Here V_{ij} and W_{ij} are the direct and exchange potentials, respectively, and in the double-channel case they have form (3.68); $V_{i\mu}^c$, $A_{\mu\nu}$ and $W_{\mu\nu}'$ are the matrix elements of the Coulomb inter-electron interaction between the correlation functions χ_μ and the functions $\Psi(\Gamma_i, X, x_{N+1})$ of the continuum, on the one hand, and between the correlation functions χ_μ and the functions of the AIS, on the other hand. Functions $U_{i\mu}$ are the matrix elements between the functions of AIS and those of the continuum.

Explicit expressions for these matrix elements V_{ij}, W_{ij}, $V_{i\mu}^c$, $U_{i\mu}$ and $A_{\mu\nu}$ in (6.39) and L_{ij} are given in [6.14], $\mathcal{E}_\mu$ is the energy of the μ-th AIS which is defined from the condition of (5.9).

Varying (6.39) with respect to F_{kl} and then adding the corresponding set of Lagrange factors M_λ in order to incorporate the condition (6.32), we obtain the system of integro-differential equations

$$\sum_j L_{ij} F_{jl} + \sum_\mu C_\mu^l V_{i\mu}^c + \sum_\nu A_\nu^l U_{i\nu} + \sum_\lambda M_\lambda P_{n_\lambda l_\lambda} \delta_{l_i l_\lambda} = 0 , \qquad (6.40)$$

while the variation of (6.39) with respect to C and A leads to the system of integro-differential equations:

$$\sum_\nu (A_{\mu\nu} - E\delta_{\mu\nu}) C_\nu^l + \sum_j \int V_{j\mu}^c F_{jl} dr + \sum_\nu W_{\mu\nu}' A_\nu^l = 0 , \qquad (6.41)$$

$$(\mathcal{E}_\mu - E) A_\mu^l + \sum_j \int U_{j\mu} F_{jl} dr + \sum_\nu W_{\mu\nu}' C_\nu^l = 0 . \qquad (6.42)$$

Using the compact matrix notation the initial system of equations which describes the electron-atom (or electron-ion) scattering can be represented in the form

$$LF + V^c C + U A + \lambda P = 0 , \qquad (6.43)$$
$$(V^c \mid F) + (A - E)C + W' A = 0 , \qquad (6.44)$$
$$(U \mid F) + (\mathcal{E} - E)A + W'C = 0 , \qquad (6.45)$$

$(P_\alpha \mid F_i) = 0$ if $l_\alpha = l_i$, and F must obey the boundary conditions (6.38).

6.3 Diagonalization Method Solution

The CCM, first introduced by Massey and Mohr as long ago as 1932, is one of the most frequently used approximation methods of solution of (6.17). It has been widely used, in particular, by *Burke* [6.17], *Burke* and *Seaton* [6.18], *Damburg* and *Gailitis* [6.19], and others. Within CCM, an approximate solution to (6.17) is looked for in the form of (6.34). The derivation of the system of equations which describe the close coupling of the channels is quite similar to the derivation of (6.43–45).

Many calculations of the electron scattering by atoms and ions have been carried out by using the CCM. We will examine the results of these calculations below. As was shown by Feshbach (Sect. 3.2), this method provides a natural description of resonances which result from the electron capture by an excited target. These calculations, however, are exceedingly complicated and time consuming.

At least three things contribute to the complexity of the CCM:

i) It is necessary to jointly solve a large number of integro-differential equations describing both open and closed channels;

ii) it is necessary to solve a system of equations with a very small energy interval in order to get an accurate picture of the shape of the resonances;

iii) an additional numerical fit of the calculated cross section must be made in order to determine the parameters of the resonances (i.e., energy and widths).

As pointed out above, the DM is free of these shortcomings. Consider now the solution of the system of equations (6.43–45) in the diagonalization approximation. Let's start with the first assumption, namely, let us neglect the interaction of the AIS with states described by the correlation functions, i.e., let W' be equal to zero. This assumption seems to be quite reasonable since W' is a matrix element of the Coulomb inter-electron interaction of the states which have weakly overlapping wave functions. Then the equations (6.44,45) are decoupled with respect to C and Λ and they can be treated as independent equations [note that these equations are implicitly coupled by means of continuum functions $F(r)$, of course]. Then the initial system of equations of the DM has the form

$$LF + V^c C + U\Lambda + \lambda P = 0 , \tag{6.46}$$

$$(V^c \mid F) + (A - E)C = 0 , \tag{6.47}$$

$$(U \mid F) + (\mathcal{E} - E)\Lambda = 0 , \qquad \text{moreover} \tag{6.48}$$

$$(P_\alpha \mid F_i) = 0 \quad \text{if} \quad l_\alpha = l_i .$$

Note that the solution of (6.47) can be written formally as

$$C = -(A - E)^{-1}(V^c \mid F) . \tag{6.49}$$

Upon substitution of (6.49) into (6.46) one eliminates the unknown coefficients C from (6.46), and the system (6.46–48) acquires the new form

$$LF - V^c(A - E)^{-1}(V^c \mid F) + U\Lambda + \lambda P = 0 , \tag{6.50}$$

$$(U \mid F) + (\mathcal{E} - E)\Lambda = 0 , \qquad \text{where} \tag{6.51}$$

$$(P_\alpha \mid F_i) = 0 \quad \text{if} \quad l_\alpha = l_i .$$

A solution F of the system (6.50) of equations is found as:

$$F(r) = F^0(r) + \int dr' G(r, r') U(r')\Lambda , \tag{6.52}$$

where $F^0(r)$ is a regular at zero solution of the system:

$$LF^0 - V^c(A - E)^{-1}(V^c \mid F^0) + \lambda P = 0 , \tag{6.53}$$

where

$$(P_\alpha \mid F_i) = 0 \quad \text{if} \quad l_\alpha = l_i , \tag{6.54}$$

with the boundary conditions

$$F^0(0) = 0 , \quad F^0(r \to \infty) \sim k^{-1/2} \left(\sin\theta + \cos\theta K^0 \right) , \tag{6.55}$$

where K is a diagonal matrix with $k_i = [2(E - \mathcal{E}_i)]^{1/2}$ elements, $\mathcal{E}_i$ is the energy of the target in the i-th channel, K^0 is the reactance matrix of the nonresonance scattering, $G(r, r')$ is the Green's matrix of the system (6.53)

$$\left(\frac{d^2}{dr^2} + \overline{W} \right) G(r, r') = \delta(r, r') , \tag{6.56}$$

where $\overline{W}$ is the integro-differential operator

$$\overline{W} = \frac{2Z}{r} - \frac{l(l+1)}{r^2} + k^2 - 2(V + W) - V^c(A - E)^{-1}V^c - \lambda P . \tag{6.57}$$

Since the nonlocal exchange potential W in the resonance region is equivalent to a certain local potential [6.20], the Green's matrix is represented as

$$G(r, r') = \begin{cases} F^0(r)\widetilde{G}^0(r') , & r < r' , \\ G^0(r)\widetilde{F}^0(r') , & r > r' , \end{cases} \tag{6.58}$$

where the tilde denotes a transposed matrix, $G^0(r)$ is the solution of the system (6.53, 54) which is irregular at zero. The function $G^0(r)$ obeys the boundary condition

$$G^0(r \to \infty) \sim -k^{1/2} \cos\theta .$$

Let us first show that the Wronskian of $F^0(r)$ and $G^0(r)$ is equal to unity, i.e.,

$$\widetilde{F}^0 G'^0 - \widetilde{F'}^0 G^0 = 1 . \tag{6.59}$$

Indeed, since $F''^0 + \overline{W} F^0 = 0$, $G''^0 + \overline{W} G^0 = 0$, $\widetilde{\overline{W}} = W$, then $\widetilde{F}^0 G''^0 - \widetilde{F''}^0 G^0 = 0$ or $\widetilde{F}^0 G'^0 - \widetilde{F'}^0 G^0 = \text{const.}$ Taking into account the boundary conditions for $F^0(r)$ and $G(r)$, we obtain (6.59). Using now (6.59) and the explicit form of the matrix $G(r, r')$, (6.58), one can show that

$$G(r, r' = r - 0) - G(r, r' = r + 0) = 0 ,$$

$$\frac{d}{dr} \left[G(r, r' = r - 0) - G(r, r' = r + 0) \right] = 1 . \tag{6.60}$$

It follows from (6.60) that $G(r, r')$ obeys (6.56). Thus, as is seen from (6.52, 58), the solution $F(r)$ of the system of equations (6.50, 51) is expressed in terms of the linearly independent solutions $F^0(r)$ and $G^0(r)$ of the system (6.53, 54).

To find Λ, we shall substitute (6.52) into (6.51). Writing the indices μ, ν explicitly, we obtain

$$\left(U_\mu \mid F^0 + \int dr' G(r, r') \sum_\nu U_\nu \Lambda_\nu \right) + \sum_\nu (\mathcal{E}_\mu - E) \delta_{\mu\nu} \Lambda_\nu = 0 , \qquad (6.61)$$

or

$$\sum_\nu \left[(E - \mathcal{E}_\mu) \delta_{\mu\nu} - (U_\mu \mid G \mid U_\nu) \right] \Lambda_\nu = (U_\mu \mid F^0) . \qquad (6.62)$$

According to [6.20], Λ can be determined by solving the system of linear inhomogeneous equations (6.62), which is a tedious task. Therefore, at this stage the DM introduces the main "diagonalization" assumption, according to which the off-diagonal terms $(U_\mu \mid G \mid U_\nu)$, $\mu \neq \nu$ in (6.62) may be neglected, i.e., the coupling of different autoionizing levels via the open channels may be neglected too. This simplification of Feshbach's method has been proposed by *Balashov* et al. in the study of the resonance photoionization in atoms [6.21]. The diagonalization approximation is the only significant simplification of this approach to the exact solution of the system (6.43–45) in comparison with Feshbach's method since the direct coupling of open and closed channels is taken into account just within the basis chosen for this group of channels. The diagonalization approximation allows one to immediately write the expression for Λ_μ

$$\Lambda_\mu = \frac{(U_\mu \mid F^0)}{E - \mathcal{E}_\mu + \overline{\Delta}_\mu} , \qquad \text{where} \qquad (6.63)$$
$$\overline{\Delta}_\mu = - (U_\mu \mid G \mid U_\nu) .$$

Upon substituting (6.63) into (6.52) we obtain

$$F(r) = F^0(r) + \int dr' G(r, r') \sum_\mu U_\mu(r') \frac{(U_\mu \mid F^0)}{E - \mathcal{E}_\mu + \overline{\Delta}_\mu} . \qquad (6.64)$$

Going in (6.64) to an asymptotics of $F(r)$ and using (6.55) as well as an asymptotics of $G^0(r)$ and comparing it with (6.38), we obtain the expression for the K-matrix of the resonance scattering in the form

$$K = K^0 + K' , \qquad \text{where} \qquad (6.65)$$

$$K' = \sum_\mu \frac{\gamma_\mu \times \gamma_\mu}{E - \mathcal{E}_\mu + \overline{\Delta}_\mu} . \qquad (6.66)$$

The elements of the vectors γ_μ are determined by

$$(\gamma_\mu)_i = \sum_j \int dr \, F_{ij}^0(r) U_{j\mu}(r) \, . \tag{6.67}$$

As is known, the K-matrix is related to the T-matrix of the scattering process by

$$T = -2iK(1 - iK)^{-1} \, . \tag{6.68}$$

The transition from the K-matrix, which contains singular terms, to the T-matrix is described in detail in [6.22]. We shall only write, therefore, the final result

$$T_{ij} = T_{ij}^0 - 2i \sum_{\lambda\mu} (\alpha_{\lambda,i}\alpha_{\mu,j}) \left[\left(e - E + \Delta - \frac{i}{2}\Gamma \right)^{-1} \right]_{\lambda\mu} \, . \tag{6.69}$$

Here T_{ij}^0 are the T^0-matrix elements of the nonresonance scattering, and the second term describes the resonance scattering and is a generalization of the Breit-Wigner formula for the multichannel case

$$\alpha_\mu = (1 - iK^0)^{-1} \gamma_\mu \, , \qquad e_{\lambda\mu} = (\mathcal{E}_\mu - \overline{\Delta}_\mu)\delta_{\lambda\mu} \, , \qquad E_{\lambda\mu} = E\delta_{\lambda\mu}$$
$$\Delta_{\lambda\mu} = (\alpha_\lambda^*, K^0\alpha_\mu) \, , \qquad \Gamma_{\lambda\mu} = 2(\alpha_\lambda^*, \alpha_\mu) \, .$$

If the distances between the energies of the neighbouring resonances are greater than the sum of their widths, then the double sum in (6.69) is reduced to the single sum [6.22]

$$T = T^0 + 2i \sum_\mu \frac{\alpha_\mu \times \alpha_\mu}{E - \mathcal{E}_\mu - \Delta_\mu + i\Gamma_\mu/2} \, , \qquad \Gamma_\mu = 2 \,|\, \alpha_\mu \,|^2 \, . \tag{6.70}$$

The differential cross section which corresponds to the transition of the target from the nL_1S_1-state into the $n'L_1'S_1'$-state has the form

$$\frac{d\sigma(nL_1S_1 \to n'L_1'S_1')}{d\Omega} = \frac{\pi}{2k_n^2(2L_1 + 1)(2S_1 + 1)}$$

$$\times \left[\delta_{nL_1S_1,n'L_1'S_1'} \left\{ \,|\, C_{n'}(\theta) \,|^2 + \pi^{-1/2} \sum_{LSl_2}(2L + 1) \right. \right.$$

$$\times \mathrm{Re} \left[-iC_{n'}(\theta)e^{-i\sigma_0}\overline{T}_{n'L_1'S_1'l_2',nL_1S_1l_2}^{LS\pi} \right] P_{l_2}(\cos\theta)$$

$$+ \sum_\lambda \frac{P_\lambda(\cos\theta)}{4\pi}(-1)^{L_1+L_1'} \sum_S(2S + 1)$$

$$\times \sum_{LP} \sum_{l_2l_2'} \sum_{p_2p_2'} Z(l_2Lp_2P; L_1\lambda)Z(l_2'Lp_2'P; L_1'\lambda) \Bigg\}$$

$$\left. \times \overline{T}_{n'L_1'S_1'l_2',nL_1S_1l_2}^{LS}\overline{T}_{n'L_1'S_1'p_2',nL_1S_1p_2}^{*LS} \right] \, , \tag{6.71}$$

where

$$C_n(\theta) = \eta_n \mathrm{cosec}^2 \frac{\theta}{2} \exp\left(-\frac{2i\eta_n \ln \sin \theta/2}{(4\pi)^{1/2}}\right) ,$$

(6.72)

$$\eta_n \equiv -\frac{Z-N}{k_n} , \quad \sigma_l = \arg \Gamma(l+1+i\mu) ,$$

$$\overline{T}^{LS\pi}_{n'L_1' S_1' l_2', nL_1 S_1 l_2} = \exp\left[i(\sigma_{l_2} + \sigma_{l_2'})\right] T^{LS}_{n'L_1' S_1' l_2', nL_1 S_1 l_2} ,$$

(6.73)

$$Z(abcd; ef) = (-1)^{(f-a+c)/2} [(2a+1)(2b+1)(2c+1)(2d+1)]^{1/2}$$
$$\times (a0c0 \mid f0)W(abcd; ef) .$$

(6.74)

Here $(a0b0 \mid f0)$ is the Klebsch-Gordan coefficient, $W(abcd; ef)$ is the Racah coefficient.

By integrating the second sum in (6.71) over angular variables we obtain the well-known expression of the total excitation cross section

$$\sigma(nl_1 S_1 \to n'L_1' S_1') = \sum_{LSl_2 l_2'} \frac{(2L+1)(2S+1)}{2k_n^2(2L_1+1)(2S_1+1)}$$

$$\times \left| \overline{T}^{LS}_{nl''L_1' S_1' l_2', nL_1 S_1 l_2} \right|^2 .$$

(6.75)

6.4 Elastic Electron Scattering
by Ions with Zero Orbital Angular Momentum
in Their Ground State

Pure elastic scattering occurs when the energy of the incident electrons is too low to cause the target excitation. Within the DM, this case corresponds to the consideration of only the first term of the expansion in (6.19), i.e., only the ground state of the target is actually involved.

Consider elastic electron scattering by ions with one valence s-electron outside of the filed subshells. Up to now, most experimental studies of electron-ion scattering were done on ions which have zero angular momentum in their ground state. These ions are He^+, Be^+, Mg^+, Ca^+, Sr^+, Ba^+, Hg^+, alkali ions and multiply charged ions of other elements. Thus, $L_1 = 0$ is an important particular case of electron scattering by a complex ion. We shall discuss this method in more detail since further specific calculations deal with ion targets of that type.

Since a single term is retained in the expansion (6.19) and since the target has $L_1 = 0$, the system (6.46) is reduced to a single equation which describes pure elastic electron scattering.

Moreover, since the ions considered have a single unfilled s-subshell, then the virtual capture of the incident electron into this subshell is possible only in the ^1S state. Therefore, the system of the algebraic equations (6.47) is also reduced to a single equation for the unknown coefficient C, only for the ^1S state. Since the single continuum ($i = j = 1$) and a single correlation function $\chi_1(\mu = 1)$ are present here, we shall omit the indices i, j and μ in the present section. Thus, the system of equations (6.46–48) is simplified significantly and may be written as

$$\left[\frac{d^2}{dr^2} - \frac{L(L+1)}{r^2} + \frac{2Z}{r} + k^2 - 2V(r)\right] F(r)$$

$$= 2 \int dr' W(r, r') F(r') + 2V^c(r)C + \lambda_{nl} P_{nl}(r) + \sum_\mu U_\mu(r)\Lambda_\mu , \qquad (6.76)$$

$$(V^c \mid F) + A - E)C = 0 , \qquad (6.77)$$

$$(U_\mu \mid F) + (\mathcal{E}_\mu - E)\Lambda_\mu = 0 , \qquad (6.78)$$

$$(P_{nl} \mid F) = 0 \quad \text{if} \quad l = L .$$

Here $n = 1$ for He$^+$, $n = 2$ for Be$^+$ and $n = 3$ for Mg$^+$. Since the wave functions of the He$^+$ target are exact solutions of the Schrödinger equation (6.31), one may write an expression for the potentials of the direct and exchange interaction $V(r)$ and $W(r, r')$, respectively, as well as that for the potential of the correlation interaction $V^c(r)$ and for the matrix element A in an explicit analytical form. Namely, taking into account that $P_{1s}(r)$ for He$^+$ is

$$P_{1s}(r) = 2Z^{3/2}r\, e^{-Zr} = 2^{5/2}r\, e^{-2r} , \qquad (6.79)$$

we obtain

$$V(r) = Y_0(P_{1s}, P_{2s} | r) = -2e^{-4r}(1 + 1/2r) + 1/r ,$$

$$W(r, r') = (-1)^S \frac{P_{1s}(r)P_{1s}(r)}{2L + 1} \frac{r_<^L}{r_>^{L+1}}$$

$$= (-1)^S \frac{32rr'\exp[-2(r + r')]}{2L + 1} \frac{r_<^L}{r_>^{L+1}} , \qquad (6.80)$$

$$V^c(r) = \left[\varepsilon_{1s} + 2Y_0(P_{1s}, P_{1s} \mid r)\right] P_{1s}(r) ,$$

$$A = 2\varepsilon_{1s} + 2 \int_0^\infty dr\, P_{1s}^2(r)Y_0(P_{1s}, P_{1s} \mid r) ,$$

where

$$Y_\lambda(P_\alpha P_\beta \mid r) = r^{-\lambda-1} \int_0^r dr'\, P_\alpha(r')P_\beta(r')r'^\lambda$$

$$+ r^\lambda \int_r^\infty dr'\, P_\alpha(r')P_\beta(r')r'^{(-\lambda-1)} .$$

Since for an H$^+$ ion $\varepsilon_{1s} = 4$ Ry, the two last expressions in (6.80) become

$$V^c(r) = 2^{7/2}r\, e^{-2r} \left\{\frac{1}{r} - 2\left[\left(\frac{1}{2r} + 1\right)e^{-4r} + 1\right]\right\} , \qquad (6.81)$$

$$A = -8 + 2 \times \tfrac{5}{8} \times 2 = -5.5 \text{ Ry} .$$

90

Hence, $A - E = A_{11} - \varepsilon_{1s} - k^2 = (1.5 + k^2)$. Since $(1.5 + k^2)^{-1}$ is nonsingular for any k^2, C can be defined from (6.77) and can be substituted into (6.76). Then the system of equations of the DM for the elastic electron-He$^+$-ion scattering cross section is

$$
\left[\frac{d^2}{dr^2} - \frac{L(L+1)}{r^2} + \frac{2}{r} + k^2 + 4e^{-4r}\left(1 + \frac{1}{2r}\right) F(r) \right]
$$

$$
= (-1)^S \frac{64 r e^{-2r}}{2L+1} \left(\frac{1}{r} \int_0^r r' e^{-2r'} F(r') dr' + \int_r^\infty \frac{e^{-2r'} F(r')}{r'} dr' \right)
$$

$$
+ 2^{9/2} r e^{-2r} \left\{ \frac{1}{r} - 2\left[1 + e^{-4r}\left(1 + \frac{1}{2r}\right)\right] \right\} \frac{1}{1.5 + k^2} \int_0^\infty 2^{7/2} r' e^{-2r'}
$$

$$
\times \left\{ \frac{1}{r'} - 2\left[1 + e^{-4r}\left(1 + \frac{1}{2r'}\right)\right] \right\} F(r') dr'
$$

$$
+ \lambda_{1s} P_{1s}(r) + \sum_\mu U_\mu(r) \Lambda_\mu , \tag{6.82}
$$

$$
(U_\mu \mid F) + (\mathcal{E}_\mu - E)\Lambda_\mu = 0 , \quad (P_{1s} \mid F) = 0 \quad \text{if} \quad L = 0 ,
$$

where

$$
U_\mu(r) = \sum_{n_3 l_3, n_4 l_4} \beta_{n_3 l_3, n_4 l_4} \left[\sum_\lambda f_\lambda(l_1 l_2 l_3 l_4; L) Y_\lambda(P_{n_1 l_1}, P_{n_3 l_3} \mid r) P_{n_4 l_4}(r) \right.
$$

$$
\left. + (-1)^S \sum_\lambda g_\lambda(l_1 l_2 l_3 l_4; L) \times Y_\lambda(P_{n_1 l_1}, P_{n_4 \, n_4} \mid r) P_{n_3 l_3}(r) \right] , \tag{6.83}
$$

The angular matrix elements F_λ and g_λ which are tabulated in [6.23] are expressed by the Wigner's $3j$ and $6j$ symbols

$$
f_\lambda(l_1 l_2 l_3 l_4; L) = (-1)^{l_1 + l_3 - L} [(2l_1 + 1)(2l_2 + 1)(2l_3 + 1)(2l_4 + 1)]^{1/2}
$$

$$
\times \begin{pmatrix} l_1 & \lambda & l_3 \\ 0 & 0 & 0 \end{pmatrix} \begin{pmatrix} l_2 & \lambda & l_4 \\ 0 & 0 & 0 \end{pmatrix} \begin{Bmatrix} l_1 & l_2 & L \\ l_4 & l_3 & \lambda \end{Bmatrix} ,
$$

$$
g_\lambda(l_1 l_2 l_3 l_4; L) = (-1)^{l_2 + l_4} [(2l_1 + 1)(2l_2 + 1)(2l_3 + 1)(2l_4 + 1)]^{1/2}
$$

$$
\times \begin{pmatrix} l_1 & \lambda & l_4 \\ 0 & 0 & 0 \end{pmatrix} \begin{pmatrix} l_3 & \lambda & l_2 \\ 0 & 0 & 0 \end{pmatrix} \begin{Bmatrix} l_1 & l_2 & L \\ l_3 & l_4 & \lambda \end{Bmatrix} .
$$

For multi-electron Be$^+$ and Mg$^+$ ions, all the matrix elements of interaction can not be written down explicitly because the radial wave functions of the target are approximate Hartree-Fock functions, which are tabulated. However, taking into account the simple structure of the wave function of the ground-state target for the ion in question, one can write the matrix elements in terms of $Y_\lambda(P_\alpha P_\beta \mid r)$-functions. For example, the direct and exchange interaction potentials for a Be$^+$ ion may be written as

$$V(r) = 2Y_0\left(P_{1s}, P_{1s} \mid r\right) + Y_0\left(P_{2s}, P_{2s} \mid r\right)\lambda,$$

$$WF(r) = W_c F(r) + (-1)^S \frac{Y_L(P_{2s}, F \mid r)}{2L+1}, \tag{6.84}$$

$$W_c F(r) = \frac{1}{2L+1}\left[Y_L\left(P_{1s}, F \mid r\right)P_{1s}(r) + Y_L\left(P_{2s}, F \mid r\right)P_{2s}(r)\right],$$

and for an Mg^+ ion

$$F(r) = 2Y_0\left(P_{1s}, P_{1s} \mid r\right) + 2Y_0\left(P_{2s}, P_{2s} \mid r\right)$$
$$+ 6Y_0\left(P_{2p}, P_{2p} \mid r\right) + Y_0\left(P_{3s}, P_{3s} \mid r\right),$$

$$WF(r) = W_c F(r) + (-1)^{1+S}\frac{Y_L(P_{3s}, F \mid r)}{2L+1}, \tag{6.85}$$

$$W_c F(r) = \frac{1}{2L+1}\left\{Y_L\left(P_{1s}, F \mid r\right)P_{1s}(r) + Y_L\left(P_{2s}, F \mid r\right)P_{2s}(r)\right.$$

$$\left. + \left[\frac{3(L+1)}{2L+3}Y_{L+1}(P_{2p}, F \mid r) + \frac{2L}{2L-1}Y_{L-1}(P_{2p}, F \mid r)\right]P_{2p}(r)\right\}.$$

To study the elastic scattering, it is convenient not to use the T-matrix which comprises a single element, but to work with a scattering phase δ which is related to T_{11} by $T_{11} = 1 - \exp(2i\delta)$. Moreover, at an energy of the incident electrons k_1^2 which corresponds to the pure elastic scattering range, the wave functions of AIS whose energies are lower than those of the excited states of the target are included. Such AIS may decay only into the elastic scattering channel, i.e., the decay of AIS may result only in the production of ground-state ions. Taking into account that in the single-channel case, $K^0 = \tan \delta_{pot}$, the product $\alpha_\mu \times \alpha_\mu$ may be written

$$\alpha_\mu \times \alpha_\mu = \frac{1 + i\tan \delta_{pot}}{1 - i\tan \delta_{pot}}\frac{\Gamma_\mu}{2} = e^{2i\delta_{pot}}\Gamma_\mu/2 \tag{6.86}$$

where δ_{pot} is a nonresonance (potential) scattering phase. Then

$$T = 1 - e^{2i\delta} = 1 - e^{2i\delta_{pot}} + ie^{2i\delta_{pot}}\sum_\mu \frac{\Gamma_\mu/2}{E - \mathcal{E}_\mu - \Delta_\mu + i\Gamma_\mu/2}. \tag{6.87}$$

It follows from (6.87) that the scattering phase δ is determined by the relation

$$e^{2i\delta} = e^{2i\delta_{pot}}\left(1 - i\sum_\mu \frac{\Gamma_\mu/2}{E - \mathcal{E}_\mu - \Delta_\mu + i\Gamma_\mu/2}\right) \tag{6.88}$$

i.e., the total scattering phase δ can be represented by a sum, $\delta = \delta_{pot} + \delta_{res}$, where δ_{res} is a resonance scattering phase shift

$$e^{2i\delta_{res}} = 1 - i\sum_\mu \frac{\Gamma_\mu/2}{E - \mathcal{E}_\mu - \Delta_\mu + i\Gamma_\mu/2}. \tag{6.89}$$

If the AIS are well-isolated in energy $\left(|\mathcal{E}_\mu - \mathcal{E}_\nu| \gg \frac{\Gamma_\mu + \Gamma_\nu}{2}\right)$, then in the vicinity of the μ-th AIS the resonance phase is

$$e^{2i\delta_{\text{res}}} = 1 - \frac{\Gamma_\mu/2}{E - \mathcal{E}_\mu - \Delta_\mu + i\Gamma_\mu/2} = \frac{E - \mathcal{E}_\mu - \Delta_\mu - i\Gamma_\mu/2}{E - \mathcal{E}_\mu - \Delta_\mu + i_\beta/2}$$

$$= \frac{\cos\delta_{\text{res}} - i\sin\delta_{\text{res}}}{\cos\delta_{\text{res}} + i\sin\delta_{\text{res}}}.$$

Then we obtain the known expression for δ_{res}

$$\delta_{\text{res}} = -\arctan\frac{\Gamma_\mu/2}{E - \mathcal{E}_\mu - \Delta_\mu}, \tag{6.90}$$

according to which the total scattering phase $\delta = \delta_{\text{pot}} + \delta_{\text{res}}$ increases by π while passing the resonance.

The shape of the elastic scattering cross section in the vicinity of the isolated AIS is now determined by the relation between the quantities δ_{pot} and δ_{res}. So, for instance, if at the resonance points $(E = \mathcal{E}_\mu + \Delta_\mu)$ $\delta_{\text{pot}} \approx 0$, then AIS will give rise to the Breit-Wigner peaks in the cross sections. If, however, $\delta_{\text{pot}} \approx \pi/2$ at the resonance points, then the AIS will produce deep minima (the so-called transparency gaps).

The differential cross section (6.71) for the elastic electron scattering by single-charged ions with a single external valence electron (s-electron) may be written in a more convenient form. We represent the cross section with a fixed spin by

$$\frac{d\sigma^S}{d\Omega} = \frac{1}{k^2}\left|\sum_L (2L+1)f^{LS}P_L(\cos\theta)\right|^2, \tag{6.91}$$

where f^{LS} is the partial amplitude of elastic scattering

$$f^{LS} = \frac{\exp(2i\delta_{\text{pot}}^{LS}) - 1}{2i} - e^{2i\delta_{\text{pot}}^{LS}}\sum_\mu \frac{\Gamma_\mu^{LS}/2}{E - \mathcal{E}_\mu^{LS} - \Delta_\mu^{LS} + i\Gamma_\mu^{LS}/2}. \tag{6.92}$$

Since δ_{pot}^{LS} is the sum of the Coulomb phase shift σ_L and the phase shifts δ_S^{LS} due to the short-range non-Coulomb potential, the substitution of (6.92) into (6.91) finally provides a formula similar to (5.66)

$$\frac{d\sigma^S}{d\Omega} = \frac{1}{4k^4 \sin^4 \theta/2}\,|1 + N^S|^2,$$

with

$$N^S = \frac{k}{i}\sin^2\frac{\theta}{2}\exp\left(-\frac{i}{k}\ln\sin^2\frac{\theta}{2}\right)\sum_L(2L+1)e^{2i(\sigma_L - \sigma_0)}$$

$$\times \left[e^{2i\delta_S^{LS}} \left(1 - 2i \sum_\mu \frac{\Gamma_\mu^{LS}/2}{E - \mathcal{E}_\mu^{LS} - \Delta_\mu^{LS} + i\Gamma_\mu^{LS}/2} \right) - 1 \right]$$
$$\times P_L(\cos\theta) \,. \tag{6.93}$$

As is seen from (6.93), the term N^S is a measure of the deviation of the interaction potential from the Coulomb form. The form of N^S suggests that the resonance will be pronounced most distinctly for the large angle scattering since just at $\theta = 180°$ a non-Coulomb part of the interaction potential gives the largest contribution.

6.5 Resonance Excitation of np $^2P^0$ and Metastable ns ^2S States of Ions with Ground State Zero Angular Momentum

Now consider in more detail the equations of the DM which describe the process of electron scattering by atomic ions with ground state zero angular momentum at electron energies sufficient to excite several low-lying ionic states. This situation is described by the DM if the expansion (6.19) includes the wave functions of all states which may be excited in the scattering process, in addition to the wave function of the ground state. In other words, the first term of the expansion (6.35) contains all the channels which are open at a given energy. In this case, as is shown in [6.24, 25], the obtained scattering eigenphases are the lower limits of their true values and subsequently increasing the number of ion states in (6.19) will lead to the monotonic increase of the scattering eigenphases, making them converge to their true values.

For example, for an He$^+$ ion, at an electron energy which is sufficient to excite the $n = 2$ levels, the 1s, 2s, and 2p states of He$^+$ must be included in the first term of (6.35). This procedure provides the system of four (three, if $L = 0$) integro-differential equations. Since the wave functions of the 1s, 2s, and 2p states of He$^+$ are exact solutions of the Schrödinger equation and are expressed in an explicit analytical form, one may write a corresponding system of equations of the DM which describes the open channels of the scattering process. An explicit forms of this system of equations is presented in [6.26]. To find solutions of this system one of the most effective numerical methods of the solution of the second-order differential equations without the first derivative (de Vodgeler's method Ref. 6.27 was used in [6.26]). The accuracy of the integration of integro-differential equations, which was better than 10^{-2} in all cases under consideration, was controlled by means of K^0-matrix symmetry.

In order to obtain the excitation cross sections for the resonance 2p and 3p states of the Be$^+$ and Mg$^+$ ions in the energy range below the 3s and 4s excitation thresholds, respectively, the abovementioned resonance np $^2P^0$ states of Be$^+$ and Mg$^+$ must be included into the first term in (6.35) too. Then the resulting system of equations will contain four (three, if $L = 0$) integro-differential equations. When the orthogonality conditions are taken into account, the corresponding correlation functions χ_μ which allow for the virtual electron capture into the unfilled valence ns and np subshells must be introduced into the expansion (6.35).

In the case of the Be^+ ion it is necessary to add the correlation functions with the following configurations: $^1S - 2s^2$ and $2p^2$, $^{1,3}P^0 - 2s2p$, $^1D - 2p^2$. For other $LS\pi$ states there are no correlation functions. The same correlation function sets must be added to the expansion while carrying out the calculation of the $3p\ ^2P^0$ excitation cross section of an Mg^+ ion too.

To solve these systems of equations, the method of transformation of the system of integro-differential equation into that of linear algebraic equations suggested by *Seaton* [6.28] was used in [6.29]. The resulting system has been solved with the help of the IMPACT routine [6.30]. The method of algebraic equations has a significant advantage over traditional methods of solution of systems of integro-differential equations, namely saving computer time and memory. This method is especially convenient in the calculations of electron scattering by complex atoms, when multi-configuration wave functions of the target are used. Note that the use of IMPACT requires no explicit form of the system of integro-differential equations to be written, since a particular form of this system is generated automatically by IMPACT depending on the type of input channels.

The multichannel character of the scattering process in the simultaneous consideration of both the elastic scattering by the ground and excited states of the ion target and of the transitions between the different states of the target requires a detailed analysis of the decay of the AIS, which are involved in the expansion (6.35), the different states of the residual ion. Therefore, in contrast to those AIS which decay via the single channel, the AIS located below the high-lying levels of the ion (e.g., below the $n = 3$ states of the He^+ ion) decay not only into the ground state of He^+ but also into the $n = 2$ levels. It will be shown below that the decay occurs mainly in this way. Therefore, the AIS which lie below the higher ionic levels are characterized, besides by the total autoionizing width Γ, by the set of the partial widts Γ_i which yield the probability of decay of the AIS into given lower ionic states. It follows from (6.70) that, with direct excitation being neglected, the differential cross section for the scattering which is accompanied by the excitation of the i-th state will take the form

$$\frac{d\sigma(1-i)}{d\Omega} = \frac{1}{k_1^2} \frac{\Gamma_1 \Gamma_i}{(E - \mathcal{E} - \Delta)^2 + \frac{1}{4}\Gamma^2} .$$

Here Γ_1 describes the probability of creation of an AIS in the input (1st) channel, Γ_i is the probability of decay of an AIS into the i-th channel. Therefore, it is convenient to use $\overline{\Gamma}_i = \Gamma_i/\Gamma$ where Γ is the total autionization probability.

In conclusion, consider briefly the expressions for the differential cross sections in the case of $ns-n's$ and $ns-n'p$ transitions. Whereas we need to obtain an explicit form of the differential cross sections for the excitation of the resonance and metastable ns levels of ions with ground state zero angular momentum, we shall consider only the second term in the sum (6.71). In the case of ns–$n's$ transitions ($L_1 = L_1' = 0$, $l_2 = l_2' = L$, $p_2 = p_2' = P$) this term becomes

$$\frac{1}{16k_1^2} \sum_{\lambda} P_{\lambda}(\cos\theta) \sum_{S}(2S+1) \sum_{LP} Z^2(LLPP;0\lambda)\overline{T}_{n'0L,n0L}^{LS}\overline{T}_{n'0P,n0P}^{*LS} \; .$$

Since

$$Z^2(LLPP;0\lambda) = (2\lambda+1)(2L+1)(2P+1)\begin{pmatrix} L & P & \lambda \\ 0 & 0 & 0 \end{pmatrix}^2 \, ,$$

then

$$\frac{d\sigma(ns-n's)}{d\Omega} = \frac{1}{16k_1^2} \sum_{S}(2S+1) \sum_{LP}(2L+1)(2P+1)$$

$$\times \overline{T}_{n'0L,n0L}^{LS}\overline{T}_{n'0P,n0P}^{*LS} \sum_{\lambda} P_{\lambda}(\cos\theta)\,(2\lambda+1)\begin{pmatrix} L & P & \lambda \\ 0 & 0 & 0 \end{pmatrix}^2 \, .$$

$$(6.94)$$

Taking into account that

$$\sum_{\lambda}(2\lambda+1)P_{\lambda}(\cos\theta)\begin{pmatrix} L & P & \lambda \\ 0 & 0 & 0 \end{pmatrix}^2 = P_L(\cos\theta)P_P(\cos\theta)$$

we obtain an expression for the differential cross section of the $ns - n's$ transition

$$\frac{d\sigma(ns-n's)}{d\Omega}$$

$$= \frac{1}{16k_1^2} \sum_{S}(2S+1)\left|\sum_{L}(2L+1)\overline{T}_{n'0L,n0L}^{LS}P_L(\cos\theta)\right|^2 \, .$$

$$(6.95)$$

The similar expression for the differential cross section for the transitions ($L_1 = 0$, $L_1' = 1$, $l_2 = L$, $l_2' = L \pm 1$, $p_2 = P$, $p_2' = P \pm 1$) is

$$\frac{d\sigma(ns-n'p)}{d} = \frac{1}{16k_1^2} \sum_{S}(2S+1) \sum_{LP}(2L+1)(2P+1)$$

$$\times \left\{ -[(2L+3)(2P+3)]^{1/2}\,\overline{T}_{n'1L+1,n0L}^{LS}\overline{T}_{n'1P+1,n0P}^{*LS} \right.$$

$$\times \sum_{\lambda}(2\lambda+1)\begin{pmatrix} L & P & \lambda \\ 0 & 0 & 0 \end{pmatrix}\begin{pmatrix} L+1 & P+1 & 1 \\ 0 & 0 & 0 \end{pmatrix}\begin{Bmatrix} L+1 & L & 1 \\ P & P+1 & \lambda \end{Bmatrix}$$

$$\times P_{\lambda}(\cos\theta) + [(2L+3)(2P-1)]^{1/2}\,\overline{T}_{n'1L+1,n0L}^{LS}$$

$$\times \overline{T}_{n'1P-1,n0P}^{*LS} \sum_{\lambda}(2\lambda+1)\begin{pmatrix} L & P & \lambda \\ 0 & 0 & 0 \end{pmatrix}\begin{pmatrix} L+1 & P-1 & \lambda \\ 0 & 0 & 0 \end{pmatrix}$$

$$\times \begin{Bmatrix} L+1 & L & 1 \\ P & P-1 & \lambda \end{Bmatrix} P_{\lambda}(\cos\theta) + [(2L-1)(2P+3)]^{1/2}$$

$$\times \, \overline{T}^{LS}_{n'1L-1,n0L} \overline{T}^{*LS}_{n'1P+1,n0P} \sum_\lambda (2\lambda+1) \begin{pmatrix} L & P & \lambda \\ 0 & 0 & 0 \end{pmatrix} \begin{pmatrix} L-1 & P+1 & \lambda \\ 0 & 0 & 0 \end{pmatrix}$$

$$\times \left\{ \begin{matrix} L-1 & L & 1 \\ P & P+1 & \lambda \end{matrix} \right\} P_\lambda(\cos\theta) - [(2L-1)(2P-1)]^{1/2}$$

$$\times \, \overline{T}^{LS}_{n'1L-1,n0L} \overline{T}^{*LS}_{n'1P-1,n0P} \sum_\lambda (2\lambda+1) \begin{pmatrix} L & P & \lambda \\ 0 & 0 & 0 \end{pmatrix}$$

$$\times \begin{pmatrix} L-1 & P-1 & \lambda \\ 0 & 0 & 0 \end{pmatrix} \left\{ \begin{matrix} L-1 & L & 1 \\ P & P-1 & \lambda \end{matrix} \right\} P_\lambda(\cos\theta) \Bigg\} . \tag{6.96}$$

Thus, the expressions which have been obtained above are the realization of an envisaged program of the dynamical description of resonances for a specific ($L_1 = 0$) case. The second addend in (6.69) is due to the Breit-Wigner resonances. The dynamical description of the resonance scattering means that the resonances are treated in terms of the initial Hamiltonian without any introduction of the arbitrary fitting parameters.

The corresponding expressions for a general case ($L_1 \neq 0$) may be obtained from the relations presented above.

In concluding this chapter we note that within the DM the diagonalization of the Hamiltonian over the subspace of closed channels is carried out using the solution of algebraic equations, whereas that over the subspcace of open channels requires solving a system of integro-differential equations.

Also widely used in the theory of electron-atom collisions is a method in which the diagonalization of the Hamiltonian in the space of closed and open channels is reduced to the solution of a system of algebraic equations. This method is called the R-matrix method and was developed by Wigner and Eisenbud in 1947 in their seminal paper [6.31] on the theory of nuclear reactions. An application of this method to the description of electron scattering by atoms has been shown in [6.32]. The R-matrix method has been used widely in calculations of electron scattering by both atoms and molecules.

The basic idea of this method is to partition the configuration space of the ($N+1$)-electron system, consisting of the atom plus the electron, into two parts: an inner part $r < a$ and an outer part $r > a$, where a is the distance from the center of the atom to the electron. The radius of the inner part a is chosen as small as possible under the restriction that all the radial wave functions $P_{nl}(r)$ of the atomic electrons must vanish with the required accuracy at $r \geq a$.

Let us examine the solution (U_i) of the system of equations of the CCM in the region $r \leq a$ under the boundary conditions

$$U_i(0) = 0 , \qquad a \frac{dU_i(r)}{dr} \bigg|_{r=a} = BU_i(a) , \tag{6.97}$$

where B is some arbitrary fixed real number (e.g., zero). It turns out that solutions of this sort exist for certain discrete values E_λ of the energy of the system. We denote these solutions by $U_{i\lambda}$. A method for finding $U_{i\lambda}$ is described in [6.32] and elsewhere. The idea is to write the functions $U_{i\lambda}$ as linear combinations of known functions U_λ^0 which satisfy the same boundary conditions as the function $U_{i\lambda}$. The functions $U_\lambda^0(r)$ describe the radial motion of an electron in a potential field $V(r)$ which serves as a model of the target field. The problem of finding $U_{i\lambda}$ then reduces to one of solving a system of algebraic equations for the expansion coefficients. It can be then shown [6.32] that the values of the functions F_{ij} and their derivatives at the boundary $r = a$ are related by

$$F_{ij}(a) = \sum_{i'} R_{ii'}(B, E) \left(a \frac{dF_{i'j}}{dr} - bF_{ij} \right)_{r=a} , \qquad (6.98)$$

where the R-matrix is

$$R_{ij}(B, E) = \frac{1}{2a} \sum_\lambda \frac{U_{i\lambda}(a)U_{j\lambda}(a)}{E_\lambda - E} . \qquad (6.99)$$

Once the R-matrix has been found, it is a simple matter to find the solution F_{ij} of the system of integro-differential equations in the outer region which satisfies the given asymptotic condition. In the outer region, all the short-range and exchange potentials are zero, so that it is a simple matter fo find a complete set of linearly independent solutions in this region. The solution being sought, F_{ij}, can then be written as a linear combination of these solutions, and the coefficients of the linear combination can be found from (6.98) and the asymptotic condition in the limit $r \to \infty$.

We have two comments here. First, although the R-matrix does depend on the particular choice of the constants a and b, observables do not depend on them. Second, the R-matrix method has the important advantage that the E dependence of the R-matrix is present only in the dominator, as can be seen from (6.99), so that it is a trivial matter to find it for various values of E once the values of $U_{i\lambda}(a)$ are known.

We might also mention some other theoretical methods, which have been used relatively infrequently to study resonance phenomena in electron collisions. Examples are the quantum-defect theory (QDT), the Random-Phase Approximation with Exchange (RPAE) and an algebraic variational method. We will not discuss these methods in any detail here; we refer the interested reader to [6.3, 33].

The QDT has been used to study series of resonances in electron-ion collisions which converge on various excited states of the cerium ion. As in the R-matrix method, configuration space is partitioned into two regions. In the outer region, the wave function of the scattered electron is taken to be a purely Coulombic wave function, while in the inner region it is constructed with the help of a quantum defect. Despite the simplicity, QDT yields fairly accurate results in cases in which the resonances lie near threshold.

The RPAE is a version of the perturbation method in many-body theory. It uses a summation of certain sequences of diagrams to calculate the amplitudes of processes in electron-atom collisions. The intermediate states in the RPAE, however, are treated as particle-hole states, so that this method can describe only shape resonances and resonances which are caused by autoionizing states in processes in which atoms are ionized by electrons. These AIS are formed in a one-electron excitation of inner shells of the atom. In order to describe the Feshbach resonances which result from two-electron excitations, it is necessary to generalize the RPAE to incorporate two-particle – two-hole states among the intermediate states. This generalization was carried out and applied to the ionization of atoms in [6.34].

In summary, several theoretical methods are available which, as we will see below, give a good quantitative description of the electron scattering by atoms and ions at low energies and can furnish accurate values of the parameters of low-lying resonances. The high-lying resonances, at energies at which there are many open channels, at present lie beyond the capabilities of existing theoretical methods.

7. Survey of the Experimental and Theoretical Results

The possibility of creating narrow monoenergetic electron beams has made it possible to resolve separate resonances in low-energy electron-atom scattering cross sections. Here we present a brief review of the present state of the experiments which serves as a guide for a reader in the virtual ocean of information on this subject.

The theoretical interpretation of the obtained results is now mainly performed using CCM, the R-matrix method, the algebraic variational method and, more recently, the DM. These methods allow one to make a dynamical description of the resonances, i.e., to calculate the energies and the widths of the AIS, the fine structure of the total and differential cross sections as well as the other characteristics of the scattering processes with no phenomenological parameters introduced. A survey of these results and comparison with the experimental data is also the subject of the present chapter. We do not try to attain a complete review of all available papers; the objective is to illustrate the possibilities of the current theory in the precise description of resonance phenomena.

7.1 Experimental Techniques

Let us briefly discuss the most common experimental methods in studies of resonances in electron scattering by atoms and ions.

A monoenergetic electron beam is scattered either as it passes through a gas in a collision chamber or as it intersects an atomic or ionic beam. There are four basic types of experiments, depending on what is detected after the scattering. In the experiments of the first type ("electric" experiments), the electrons are detected. In experiments of the second ("optical") type, the radiation from the atom or ion excited as a result of the collision is detected. If the lifetime of the excited atom or ion is longer than the time required for the atom or ion to travel to the detector, one can resort to experiments of a third type: detecting the ions or atoms in the metastable excited state (the "metastable spectroscopy" method). The experiments of the fourth type use a coincidence technique. In these experiments, various pairs of particles which are products of the collision may be selected as the particles to be detected.

In most cases, the resonances have a width of 10^{-1}–10^{-3} eV or less. In order to detect a resonance, it is thus necessary to produce an electron beam in which

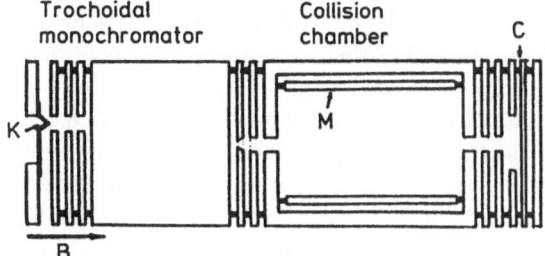

Trochoidal
monochromator

Collision
chamber

C

K

M

B

Fig. 7.1. Layout of an electron transmission experiment showing the trochoidal electron monochromator to
reduce the electron beam energy
spread. From [7.1]

the energy spread is of the same order of magnitude or less. The quantity usually
adopted as a measure of the energy spread of the beam is ΔE, the Full Width
of the electron energy distribution at Half-Maximum (FWHM). We will refer to
ΔE as simply the "energy spread" of the beam.

The Pierce electron guns which are ordinarily used provide an energy spread
~ 0.3 eV as best. In order to achieve a narrower energy distribution, one should
therefore use electron velocity selection. Monochromators with various electric
and magnetic field configurations are used for this purpose. The monochromators
which are presently used most extensively are the trochoidal electron monochromators (Fig. 7.1), which operate in a longitudinal magnetic field of $100 - 200$ Oe,
and electrostatic monochromators, a 127° cylindrical monochromator (Fig. 7.2)
and a 180° hemispherical monochromator. In the latter two monochromators
the electric field is produced between cylindrical or spherical surfaces. These
monochromators usually provide beams with an energy spread of ~ 20–40 meV.

Monoenergetic electrons can also be produced by the photoionization of
atoms by monochromatic light. As far back as 1983 *Stumpt* and *Gallagher* [7.3]
used this approach to produce an electron beam with $\Delta E = 7$ meV, which they
used to study resonances in electron scattering by inert gas atoms. There is rea-

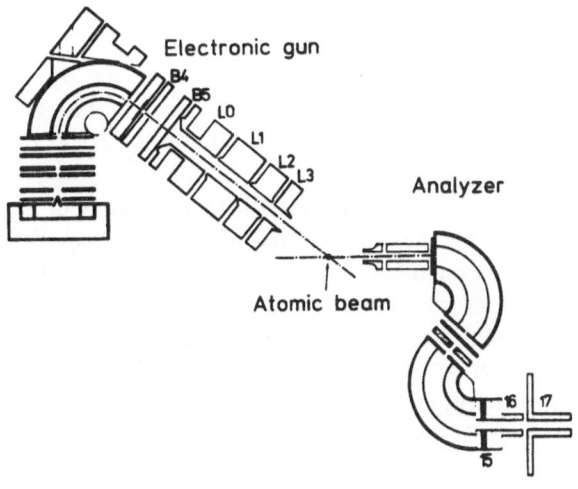

Electronic gun

B4
B5
L0
L1
L2
L3

Analyzer

Atomic beam

16   17

15

Fig. 7.2. Layout of *Eyb* and *Hofmann's* spectrometer showing the
127° electrostatic electron beam
monochromator. From [7.2]

101

son to believe that this method could be developed further to achieve an energy spread $\Delta E \sim 1$ meV. It should be noted that such electron beams have quite low intensity.

Another important characteristic of an electron beam is its current, I_0. As this current is raised (at a given electron energy), the number of electron-atom collisions per unit time increases, as does the ratio of the useful signal to the background. On the other hand, an increase in the current also increases the space charge density in the beam, which in turn degrades the energy spread ΔE. Cylindrical monochromators can furnish a beam with $\Delta E \sim 20$ meV at a current $I_0 \sim 10^{-8}$ A. Special improvements allow one to obtain beams having an energy spread of 10 meV but the current then is only about 0.1 nA [7.4]. *Kennerly* et al. [7.5] have obtained an energy spread of about 2 meV, using photoionization but the current was extremely small (~ 0.01 nA).

Experiments of the electric type have been carried out in various ways, depending on which cross section is to be measured. The total cross section for the interaction of electrons with atoms is measured in a "transmission" experiment. Figure 7.1 shows the arrangement for an experiment of this type, which was refined by *Sanche* and *Schulz* [7.1]. A beam of electrons emitted by the cathode K goes to the input of the trochoidal monochromator, within which the longitudinal magnetic field B performs a velocity selection on the electrons. At a current of 5×10^{-9} A the FWHM is 30 meV. After the electrons have been accelerated to the desired energy, they are sent into a collision chamber which is filled with a gas to a pressure of 10^{-2} Torr. Those electrons which pass through the gas and are not scattered strike collector C. Those electrons which lose kinetic energy in the axial direction as a result of collisions are retarded and reflected by the potential of an electrode in front of the collector. Also incident on the collector, of course, are electrons which have undergone forward elastic scattering, but they can be ignored if the geometry is chosen appropriately. The current I drawn by the collector is then related to the current at the entrance to the collision chamber, I_0, by

$$I = I_0 \exp(-\varrho l \sigma) , \tag{7.1}$$

where σ is the total interaction cross section, l is the distance traversed by the electron in the collision chamber, and ϱ is the number of atoms per unit volume. Relation (7.1) can thus be used to find σ from the measured current I and the known values of I_0, ϱ, and l.

The sensitivity of the experiment is improved by modulating the energy of the electron beam by a small alternating voltage of 0.005–0.06 V, which is applied to the cylinder M inside the collision chamber. The scattering signal, which is proportional to the derivative of the cross section with respect to the energy is filtered out at the modulation frequency by a lock-in detector.

Since $(dI/dE)I^{-1} = -\varrho l \sigma (d\sigma/dE)\sigma^{-1}$ at $\varrho l \sigma \gg 1$, the measurement of the derivative of the current over the energy enables one to distinguish effectively the changes in the cross sections caused by the resonances from the slow varying background.

Experiments carried out to measure the total cross section do not provide information which would be of assistance in classifying the resonances, i.e., in determining the configuration, the total angular momentum L, and the total spin S of the AIS. Information of this sort is provided by differential cross sections for some type of elastic or inelastic scattering. For example, if the atom has a zero orbital angular momentum in its initial and final states, the total angular momentum L is equal to the angular momentum of the scattered electron, l, and the angular dependence of the differential cross section at the resonance energy is determined primarily by the angular dependence of Legendre polynomial, P_l. The P-resonance, for example, is seen most clearly at the scattering angles $\theta = 0°$ and $180°$ and is not observed at all at $\theta = 90°$, since $P_1(\cos \theta) = \cos \theta$.

In differential cross section measurements electron spectrometers which are a combination of an electron monochromator and analyzer are used to detect the intensity of the scattered current and its energy. Electrons which are scattered through a given angle in the collision chamber, and which have the energy to which the analyzer is tuned (that is, its electric or magnetic field), pass through and are detected.

Figure 7.2 shows the typical spectrometer layout which Eyb and Hofmann used to study the differential scattering of electrons by alkali atoms. The electrons from the electron gun are focused by a system of lenses onto the entrance of a $127°$ cylindrical monochromator. After the energy selection, the electrons are focused onto an atomic beam by the lens system $B_4 - L_3$. The double $127°$ cylindrical analyzer can be rotated around the atomic beam between $-100°$ and $+150°$. The analyzer should pass those electrons which are scattered through a certain angle and which have a certain energy. The analyzer is accordingly put in the appropriate position, and the scattered electrons are accelerated to the energy which will be passed by the analyzer and focused by a system of lenses onto its entrance. Those electrons which pass through are again accelerated and focused onto an electron multiplyer by a lens (at the bottom right in Fig. 7.2). The angular resolution of the spectrometer is $\pm 1.5°$, while the energy resolution is about 70 meV.

At a fixed energy of the incident electrons and at a fixed scattering angle, it is possible to measure the dependence of the number of scattered electrons on the energy which they lost in collisions; i.e., it is possible to measure a loss spectrum.

The electric methods in which the changes which occur in the electrons colliding with the target can be used to study only low-lying resonances, since the resolution of the best electron spectrometers is ≈ 20 meV, and the resonances become progressively more closely spaced as the energy increases. Optical methods are free of this shortcoming.

An optical method involves studying the optical excitation functions of the spectral lines of the radiation from an atom. These functions are measured with high resolution in terms of the energy of the incident electrons, achieved through the use of the same monochromators as in the electrical methods. As a rule, the radiation from the collision chamber is observed at right angles with respect to the electron beam. A system of lenses then directs the radiation to a spectral device which selects the radiation at the wavelength of interest.

We turn now to electron-ion scattering. The most reliable experimental data on the scattering of electrons by ions are obtained in experiments of optical type involving intersecting electron and ion beams. However, several difficulties confront efforts to determine experimentally the excitation cross sections in beam experiments. The greatest difficulty is the mutual effect of the space charge of the beams. Furthermore, since the ion density in the beam is usually $\sim 10^{-6}$–$10^{-7} \mathrm{cm}^{-3}$, which is well below the residual gas density in the collision chamber, the yield of reaction products (usually, photons) must be detected on top of a significant background signal, whose level is sometimes two orders of magnitude greater than the useful signal. It thus becomes necessary to maintain high vacuum ($\sim 10^{-8}$ Torr) in the collision chamber along with a continuous influx of working medium.

Figure 7.3 shows the layout of an experiment carried out to study the cross sections for the electron-impact excitation of alkaline-earth ions [7.6, 7].

Beams of positive ions are produced by a source in which atoms are ionized in two ways. The first method is a surface method, in which atoms from heated reservoir 1 diffuse into chamber 2 and are ionized on the surface of a hot cathode. In the second method, the ions are formed in the arc of a low-voltage discharge between the cathode and the front wall of the ionization chamber. The density of the ion beam can be made as high as 6×10^{-5} A/cm^2.

The ions which are produced are extracted from the ionization chamber, accelerated, and formed into a beam by the ion optics system 3, which consists of three lenses. The ions then undergo an energy selection and are separated from the atoms in 90° electrostatic capacitor 4. Part of the outer plate of the latter is made of a tungsten grid, which allows the neutral atoms, which diffuse out of the

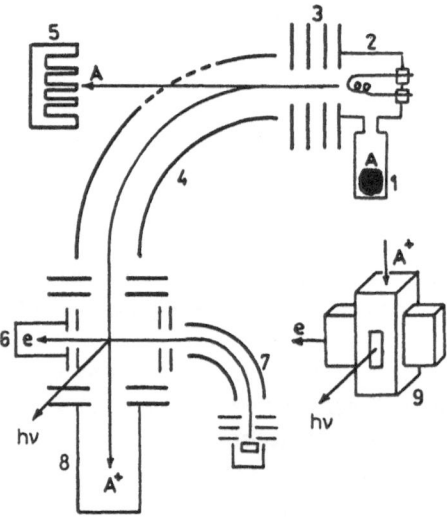

Fig. 7.3. Layout of an experiment with intersecting beams of ions and electrons. (1) A is the substance to be ionized, (2) is the ionization filament, (3) ion focusing lenses, (4) a mass and energy selector, (5) neutral beam catcher; (6) electron counter, (7) electron energy analyzer, (8) ion detector, (9) is a detail of the collision cell. From [7.6]

source, to leave the beam and be captured by a trap 5. In the collision chamber, the ion beam intersects the electron beam and then goes to a Faraday cup 8.

A beam of monoenergetic electrons is produced and selected by a 90° cylindrical electrostatic monochromator 7 and after passing the collision cell then strikes collector 6. The current density of electrons with energies of 2–20 eV ranges from 2.5×10^{-5} A/cm^2 to 6.0×10^{-4} A/cm^2. The energy spread of the beam varies over 100–300 meV. The cross sections of the ion and electron beams are 6.25 and 0.3 mm^2, respectively.

The radiation from the collision region is observed, at an angle of 90° to the collision plane, by a monochromator with a diffraction grating. This radiation is detected by a photomultiplier operated in the photon-counting mode. The useful signal is discriminated from the background which can be due to excitation of ion states in collision of ions with neutral atoms of the residual gas and of the working medium, by a modulating both beams which are out of phase by $\frac{1}{4}$ of a period. The pulsed signal is detected in two counting channels, which are synchronized with the modulation pulses.

The energy spread which has been achieved to date for the ion and electron beams is not sufficient for detecting individual resonances in electron-ion collisions. The features which are observed in the cross sections are actually the average contribution of a series of resonances.

Note that ion beams can contain long-lived excited positive ions. This difficulty can be avoided in ion trap experiments in which the metastable ions are trapped long enough so that they decay. Wolfgang Paul (from Bonn University) and Hans Dehmelt (from Washington University) have made an outstanding contribution to the ion trap techniques which earned them the *Nobel Prize* on physics in 1989.

The method of metastable spectroscopy is used to measure both differential and total cross sections for the excitation to metastable states. The differential excitation cross sections for metastable states of inert gases were measured using this method in [7.8]. In these experiments, an intense atomic beam ($\sim 10^{15}$ atoms/cm^2 s), 2 mm in diameter, was produced by a gas dynamical source with supersonic nozzle. The angular divergence of this beam was about 1 deg, and the energy spread was 60 meV for He and 50–80 meV for Ne, Ar, Kr, and Xe. The atomic beam was crossed at a right angle by an electron beam (FWHM < 80 meV) which was formed by an electrostatic 127° monochromator. The metastable atoms resulting from the collisions with electrons were deflected with respect to their initial direction of motion producing a wide angular distribution. This distribution was measured by means of a channeltron placed at different observation angles. At each chosen electron energy the angular dependence of the metastable yield was measured. Comprehensive review of results obtained by means of metastable spectrosopy is presented in [7.9].

The coincidence technique has been used most extensively in so-called (e, 2e) experiments on a electron-impact ionization of atoms. In these experiments, the scattered and ejected electrons are detected in coincidence. The measured energies and directions of motion of the electrons completely determine the kinematics of the process. *Balashov* et al. [7.10] have shown how measurements of this type can be used to obtain information on ionization through the excitation of AIS of an atom and on the symmetry of the AIS themselves.

Coincidence experiments have recenty been extended to the study of the excitation of atoms by electrons. In these so-called (e, e'γ) experiments, both the electron which excites the atom and which is scattered into a certain angle and the photon emitted in a certain direction by the excited atom are detected. Experiments of this sort require an apparatus with high time resolution (on the order of 1 ns). The requirements of a small energy spread of the electron beam naturally remain in a study of resonances. Photon-electron coincidence experiments make it possible to study the amplitudes for the excitation of sublevels of a given level which are degenerate in the magnetic quantum number (so-called coherent excitation).

Another rapidly developing technique is that of experiments with polarized beams of atoms and electrons. The information obtained from such experiments turns out to be very useful in classifiying resonances. We will discuss polarization experiments in more detail in Sect. 7.8.

In concluding this section we note that most of the experimental effort has been aimed at determining total cross sections. So far, very little work has been carried out determining differential cross sections or using polarized beams.

7.2 Hydrogen Atom

7.2.1 Elastic Scattering

The hydrogen atom is the darling of the theoreticians. Unfortunately, scattering experiments involving hydrogen atoms are very difficult because of the complexity of producing atomic hydrogen. It is not surprising then that the scattering of electrons by hydrogen atoms was studied in great detail theoretically before the corresponding experiments were carried out. However, it was only in 1962, in calculations carried out using CCM by *Smith* et al. [7.11], that it was found that the total elastic cross section increases very sharply below the excitation threshold for the $n = 2$ level of the H atom, because of a rapid increase in the ^{1}S and ^{3}P scattering phase shifts. Similar results were obtained nearly simultaneously by *Burke* and *Shey* [7.12], who worked from a detailed study of the behaviour of the scattering phase shifts near the threshold to predict the existence of a ^{1}S resonance at an energy $E = 9.61$ eV with width $\Gamma = 0.109$ eV. Burke et al. subsequently refined the values of the parameters of this resonance; in [7.13] they found $E = 9.56$ eV and $\Gamma = 0.0474$ eV, which are the values generally accepted today.

Gailitis and *Damburg* [7.14] proposed an explanation for the mechanism for the formation of resonances in cross sections for the scattering of electrons by hydrogen atoms. They started from the position that although an atom in a state with a definite parity does not have a dipole moment there can be states with a definite energy and an indefinite parity, with a nonzero dipole moment, in a case in which degenerate levels exist. An example of such a state might be the $(2s + 2p)/2^{1/2}$-state of the H atom. The attractive potential of a hydrogen atom in this state has an α/r^2 asymptotic behaviour, see (4.23). Gailitis and Damburg showed that in the case of hydrogen the value of α would be such that in the field α/r^2 there could be a set of bound states, which convert into the AIS when the coupling with open channels is taken into account. In this case the system of levels of the H^- ions would be an infinite series of doubly excited states. Each series would be characterized by a certain value of a principal quantum number n of that level of the H atom to which the given series converged and by the several possible values of the total orbital angular momentum at the given n. Each level of the series is therefore characterized by the quantum numbers $n = 2$, $L = 0 - 2$; $n = 3$, $L = 0 - 4$; $n = 4$, $L = 0 - 6$; etc. The energies of the levels of each series form a geometric progression.

The first experimental confirmation of the existence of a resonance near 9.6 eV was found by *Schulz* [7.15] in a 1964 transmission experiment in which an electron beam with an energy below 10.2 eV passed through partially dissociated hydrogen. Schulz found the resonance energy $E = 9.77 \pm 0.15$ eV, but the poor energy resolution of the electron beam (0.3 eV) prevented a determination of the width of the resonance. In subsequent experiments carried out by *Kleinpoppen* and *Raible* [7.16] and also by *Ormonde* et al. [7.17], with electron beams with a better energy resolution (~ 0.08 eV), achieved through the use of a 127° electrostatic selector, a structure in the cross section near 9.6 eV was also observed. Results which have been obtained in [7.16] are shown in Fig. 7.4. The energy of the ^1S resonance was estimated to be 9.73 ± 0.12 eV.

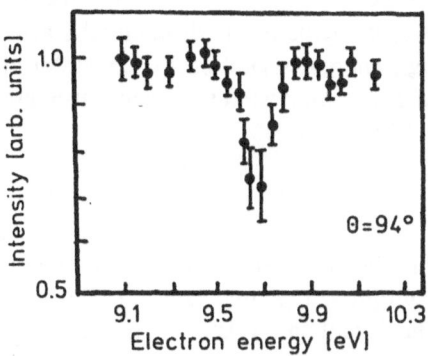

Fig. 7.4. Energy dependence of the differential cross section of e + H elastic scattering at an angle of 94°. From [7.16]

Later, some additional structure was observed in the elastic cross section below the $n = 2$ threshold of the H atom in [7.18]. The results found here were analyzed in [7.17] by the CCM. The differential cross section for elastic scattering at $\theta = 90°$ was calculated; the low-lying ^1S and ^1D resonances were taken into account. In order to compare with experimental data of [7.18] the calculated cross section was averaged over an angle of 15° and then integrated

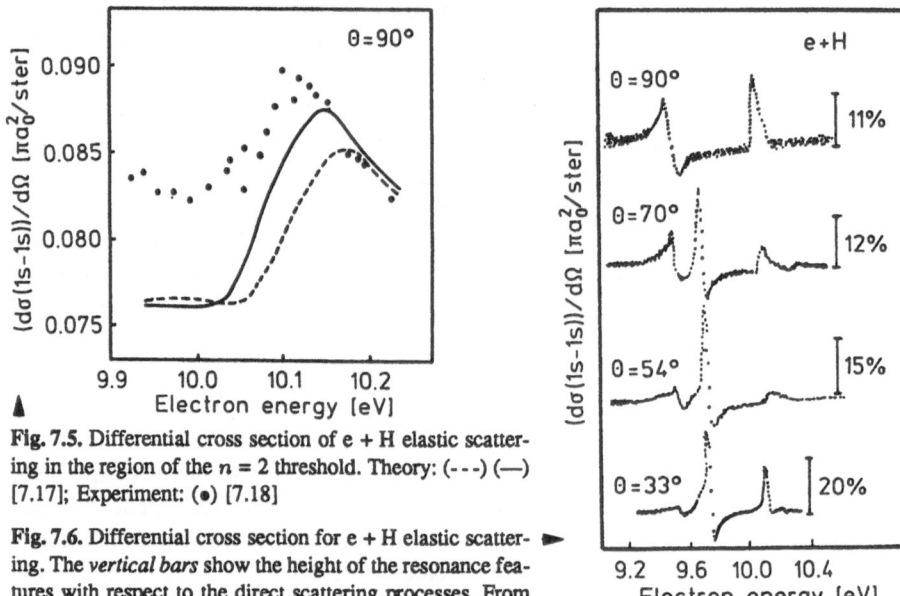

Fig. 7.5. Differential cross section of e + H elastic scattering in the region of the $n = 2$ threshold. Theory: (- - -) (——) [7.17]; Experiment: (•) [7.18]

Fig. 7.6. Differential cross section for e + H elastic scattering. The *vertical bars* show the height of the resonance features with respect to the direct scattering processes. From [7.19]

with an electron energy distribution function of 0.08 eV (FWHM). Since the cross section in [7.18] was measured in arbitrary units it was normalized in the vicinity of the lower ^1S resonance ($E = 9.56$ eV) to the calculated cross section value (Fig. 7.5). One can see from Fig. 7.5 that near the $n = 2$ threshold fairly good agreement between the theory and experiment is obtained both in the shape and in the magnitude of the cross section. The results of these calculations showed that the ^1D resonance ($E = 10.126$ eV, $\Gamma = 0.0088$ eV) was responsible for the peak below the $n = 2$ threshold. Recent measurements of the differential elastic cross section with an electron beam of high energy resolution (~ 25 meV) at energies < 10.2 eV confirmed the presence of a complicated resonance structure due to the ^1S, ^1D, and ^3P resonances (Fig. 7.6) [7.19].

7.2.2 Inelastic Scattering

More accessible from the experimental point of view is a study of the excitation of the hydrogen atom since then experiments of optical type may be used. Thus, the resonances in excitation cross sections for the 2s and 2p levels of the H atom have been the subject of a large number of experimental and theoretical investigations [7.20–25]. The cross sections are non-zero at the threshold, and this is due to the long-range potential with α/r^2 asymptotics. The presence of a shape resonance is also a typical feature of the σ (1s – 2s) and σ (1s – 2p) cross sections. The energy and the width of these resonances were calculated quite accurately in [7.21] in which the values of $E = 10.2207$ eV and $\Gamma = 0.02$ eV were obtained. A detailed comparison of theory and experiment has become possible, however, only in the mid 1970s when an electron energy beam spread $\Delta E \sim 10$ meV could be

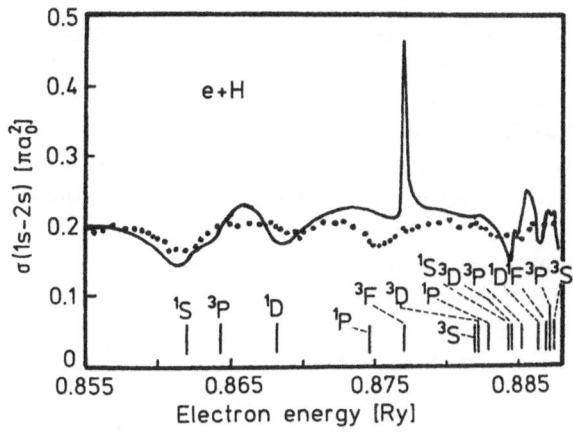

Fig. 7.7. Total cross section for electron-impact excitation of the 2s level of H. (•) experiment [7.26a], (—) calculation [7.21]

achieved. Figure 7.7 presents the calculated and measured cross sections for the 2s excitation of the H atom. The absolute cross sections were measured within a relative error of 15–17 %. Even though the calculated cross section (Fig. 7.7) has not been averaged over the energy spread in the beam, one can notice a qualitative agreement between the calculated and measured cross sections. The vertical lines mark the calculated positions of resonances.

The abundance of theoretical results in comparison to experimental data is a challenge to experimentalists. Indeed, recently *Williams* [7.26] performed a remarkably precise experiment on the determination of the total excitation cross sections for the 2s and 2p-levels of atomic hydrogen. High energy resolution, estimated to be typically 9 meV, was achieved. The absolute 2s and 2p excitation cross sections are shown in Fig. 7.8 as a function of the incident electron energy. Similarly to the previous figure, the solid lines are theoretical data of *Callaway* [7.21], convoluted with the instrumental energy resolution function. Calculations were carried out using an 18-state variational method.

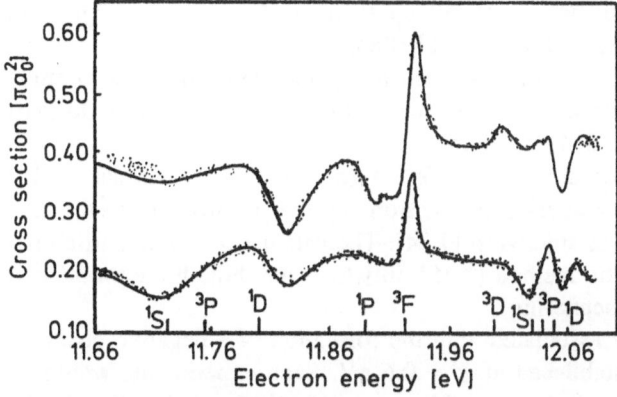

Fig. 7.8. Total cross sections for electron-impact excitation of the 2s and 2p levels of H. The upper data are the 2p cross section. *Vertical lines* indicate the positions of resonances, from [7.26b]. *Solid lines*, calculations from [7.21]

The following parameters for the ^{1}P shape resonance were also obtained experimentally: E = 10.217 eV, Γ = 22 meV for the 2p-excitation and E = 10.215 eV, Γ = 21 meV for the 2s-excitation. Excellent agreement between theory and experiment testifies to the reliability of both.

The fact that the ^{3}F resonance is dominant, which is observed also for the case of electron scattering by certain ions (Sects. 7.4–6), is a notable feature of the cross section obtained.

A high resolution electron-impact experiment was also carried out by *Rutter* et al. [7.27] to study the resonance structure in the 2s and 2p excitation cross sections for atomic hydrogen in the 11.4–12.3 eV energy interval. The resolution was estimated to be ~ 20 meV. The agreement with theoretical data of *Callaway* [7.21] is excellent. In this case the ^{3}F state is clearly pronounced.

As far as we know the only detailed calculations of electron scattering by a hydrogen atom using the HSC-method were carried out by Kuppermann and his group. That, the total, and the differential cross sections for excitation of the 2p state were calculated in [7.28]. Fairly good agreement with other calculations was obtained, and the ^{3}F resonance was also found to be dominant.

The experimental difficulties in working with atomic hydrogen have yet to be completely overcome. So far, the extensive theoretical calculations of the resonances above 12.07 eV (i.e., both below the n = 3 threshold and above it) have not received detailed quantitative experimental confirmation because of the low energy resolution of the electron beams. *Callaway* [7.29] carried out a detailed theoretical study of e-H scattering between the n = 3 and n = 4 thresholds (from 12.1 to 12.75 eV). The energies and the widths of 14 resonances were also reported in this paper. *Ho* and *Callaway* [7.30] also calculated the energy positions and the widths of resonances associated with the n = 4, 5, 6 states of the hydrogen using the complex coordinate rotation method. *Pathak* et al. [7.31] performed similar calculations using the 15-state R-matrix approach, while *Nicolaides* and *Komninos* [7.32] used a multi-configuration Hartree-Fock method and *Fukuda* et al. [7.25] utilized the HSC-method to describe high lying doubly exited states of the hydrogen. Experimental data available to check those predictions made have not been reported up to now. An electron beam with 1 meV energy resolution would be desirable here.

It is not difficult to realize that one can apply the same methods to study positron scattering by H atoms. it was shown in [7.33] that resonances do exist below the hydrogen n = 2, 3, 4 threshold.

An "electron + positronium" system (e^- + (e^+e^-)) is quite similar to that considered above. Elastic scattering of electrons by the positronium in the E = 5.105 eV energy region was studied in [7.34]. The calculations were carried out on the basis of the adiabatic method [7.35], in which the three-body problem is reduced to multi-channel scattering.

A resonance evidently associated with the formation of a negative positronium ion ($e^-e^-e^+$) was established at E = 0.6 eV. Such experiments would be of fundamental interest since they provide a test of charge conjugation in the "$e^+e^+e^-$" leptonic system. As was emphasized by the authors of [7.34], elastic scattering of slow electrons by positronium is itself of interest in view of experiments on the scattering of positronium in gases, where the existence of a large electron or positron scattering cross section might have an important effect. The same method may be used for the investigation of various low-energy scattering processes in mesoatomic physics.

7.3 Helium Atom

7.3.1 Elastic Scattering

Data on resonances in the collisions of electrons with helium atoms are of particular interest both for theoretists and experimentalists. First, the electron-atom correlation effects are revealed most clearly in doubly (and triply)-excited states of the He$^-$ negative ion. Second, the He atom is excellently suited to experimental study. These data are important for cosmology, astrophysics, and for some laboratory applications.

Around 1962, *Schulz* [7.36] carried out studies of the elastic electron scattering by helium atoms. His goal was to investigate the so-called Wigner cusps, i.e., the changes in slope in the elastic cross section at energies corresponding to the thresholds for inelastic processes. Since the first inelastic channel opens up at 19.8 eV, Schulz focused on specifically this energy region. He in fact observed a broad dip in the differential cross section for elastic scattering through 72° at this energy. However, after he more carefully calibrated his energy scale, it turned out that the structural feature which was observed occurred at 19.3 eV, i.e., 0.5 eV below the 2^3S threshold, so that it could not be a Wigner cusp. The results of these measurements are presented in Fig. 7.9. A subsequent theoretical analysis showed that this structural feature was a resonance due to a quasibound

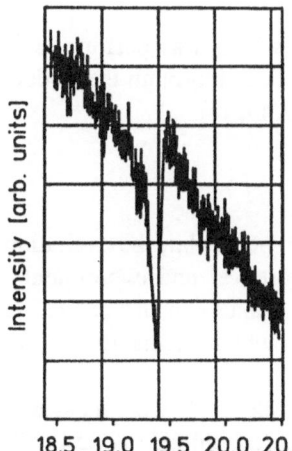

18.5 19.0 19.5 20.0 20.5
Electron energy [eV]

Fig. 7.9. Results of the first observation of the 1s2s^2 ^{2}S resonance in helium obtained by Schulz in a transmission experiment. From [7.36]

Table 7.1. Energy and width of the $1s2s^2\ ^2S$ state resonance of He^-

	Theory			Experiment			
	[7.38] a	[7.39, 49] a	[7.40] b	[7.41]	[7.42]	[7.5]	[7.44]
E [eV]	19.38	19.37	19.376	19.35 ± 0.02	19.367 ± 0.009	19.37	19.30 ± 0.01
Γ [meV]	15.1	11.72	11.56	13.0	9.0 ± 1	9.0	8 ± 2

a – R-matrix; b – Complex rotation method

state of the negative He^- ion with a $1s2s^2\ ^2S$ configuration. A sharp intensity dip is clearly seen in Fig. 7.9 showing the destructive interference of potential scattering with resonance scattering. Interferences of this type take place when the potential scattering phase shift is large and comparable with the resonance scattering phase shift and in the region below the resonance these two terms have opposite signs.

A short time later Schulz' experiments were repeated by *Simpson* and co-workers [7.37] using the transmission method. Here, constructive interference takes place and the cross section obtained by Simpson has a peak near 19.3 eV. Since then, several groups have carried out precise measurements to determine the $1s2s^2\ ^2S$ resonance. The results of these experiments are listed in Table 7.1.

The most accurate theoretical calculations on the elastic scattering of e + He were carried out by the *R*-matrix method by *Burke* et al. [7.39], who took into account eleven states of He (1^1S, $2^{1,3}S$, $2^{1,3}P^0$, $3^{1,3}S$, $3^{1,3}P^0$, and $3^{1,3}D$). As can be seen from Table 7.1, the theoretical results are in excellent agreement with the experimental values. Subsequent experiments were carried out to detect other resonances caused in elastic e + He scattering by high-lying autoionizing states which converge to the $n = 2$ threshold of He. The most detailed data were obtained in this energy interval by *Golden* [7.41], who observed thirteen resonance features in the energy range 19.3–21.3 eV (Fig. 7.10), five of which lie below the 2^3S threshold.

Although the experiment by *Andrick* et al. [7.45] which was performed at practically the same time did not reveal such structure, later thorough theoretical and experimental studies have confirmed the data of [7.43, 44].

7.3.2 Inelastic Scattering

The inelastic scattering of electrons by helium atoms, in particular, the excitation of $2^{1,3}S$ and $2^{1,3}P^0$-levels of He, is also of much theoretical and experimental interest. Several studies have been carried out on resonances in the excitation cross sections. These studies have revealed structure which stems from both closed-channel resonances and shape resonances.

The excitation functions of the $3,4,5^{1,3}S$, and 6^1S levels of He were measured in [7.46] at energies from 22.5 eV up to the ionization threshold ($E = 24.54$ eV). These measurements revealed $1sns^2\ ^2S$, $nsnp^2\ ^2P$, and $1snp^2\ ^2D$ and ^2S reso-

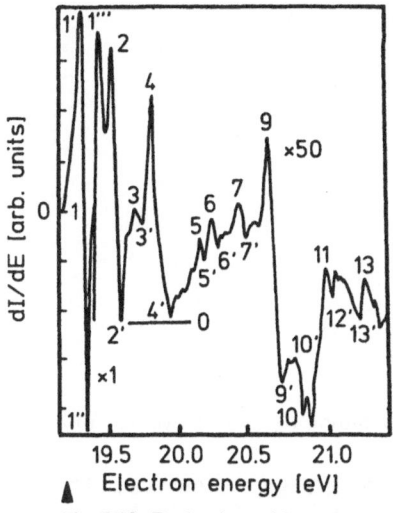

Fig. 7.10. Derivative with respect to the energy of the current of electrons passed through helium as a function of the electron energy [7.44]

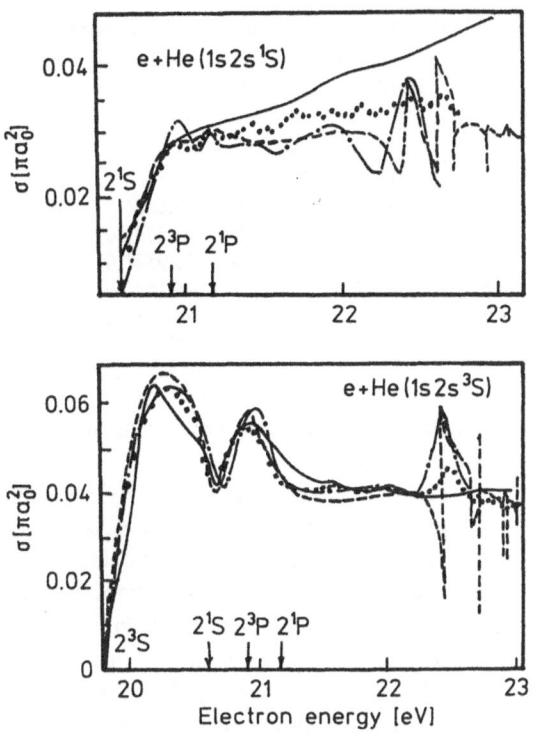

Fig. 7.11. Comparison of theoretical and experimental cross sections for electron-impact excitation of the 1s2s ^{1,3}S levels of He. (—) calculations [7.38], (– · –) calculations [7.48], (- - -) calculations [7.49], (•) experiment [7.47]

nances. Some high-resolution experiments were carried out on the excitation of the 2^3S and 2^1S-levels at energies from 19.8 to 22.7 eV in [7.47]. The results of those experiments are shown in Fig. 7.11, where they are compared with the most accurate theoretical calculations. Experimental data are presented in arbitrary units and are normalized to the magnitude of the first peak of He(2^3S), thus the good agreement with theoretical data is not accidental.

Recently a series of high-resolution experiments [7.50–52] have been done to study the resonances in the excitation functions of the $2^{1,3}$S, 3^3S, 3^3D, e^3S-states of He which are due to high-lying AIS of He$^-$, with both excited electrons having large principal quantum numbers. Table 7.2 presents the energies of resonances which were discovered experimentally in [7.50]. These AIS can be divided into distinct series corresponding to a given $n = \min(n_1, n_2)$.

The lowest AIS of each series with $n_1 = n_2$ (so-called intra-shell-type resonances) form Rydberg-type series since each electron moves in the screened Coulomb-type field of a singly charged He$^+$ ion. These AIS may be described by *Wannier's* theory [7.53–55] which has been developed to describe the two-electron ionization of atoms. Wannier's theory leads to the conclusion that the widths of the two-electron resonance levels behave like $\Gamma_n \sim n^{-3-2m} = n^{-5.254}$, where $m = 1.127$ is the threshold index for double escape from He$^-$ [7.55]. Re-

Table 7.2. Energies [eV] of the double Rydberg resonances of He [7.50]

n	$1sns^2$ ^2S	$1snsnp$ $^2P^0$	$1snsnd$ ^2D	$1sns\bar{s}$ ^2S
2	19.367(5)			
3	22.450(5)	22.600(10)	22.660(10)	22.881(5)
4	23.443(5)	23.518(10)	23.579(10)	23.667(5)
5	23.860(5)	23.907(10)	23.952(10)	23.983(10)
6	24.088(10)		24.144(15)	24.176(10)
7	24.216(10)		24.261(15)	24.288(10)

cent experimental data are, unfortunately, unable either to confirm or to refute this behaviour.

Those AIS for which $n_1 \neq n_2$ are called inter-shell-type resonances. They are produced by attachment of a very weakly bound electron by the polarization potential, e.g., of the neutral 1s3s ^{1}S-state. The first three columns in Table 7.2 contain data on intra-shell-type resonances whereas the fourth column contains the energies of inter-shell-type resonances. These data have been confirmed recently and are in agreement with the results of an experiment done with improved energy resolution [7.56]. As for the theoretical calculations of energies of high-lying AIS, the exact quantum mechanical calculations for $n = 3 - 10$ were carried out by *Komninos* et al. [7.57]. The agreement with experimental data is remarkably good.

Simple considerations allow an estimation of the energies of AIS in a system of two valence electrons in the Coulomb field of the nucleus for the case of intra-shell resonances. Due to the inter-electron repulsion, the nucleus and the electrons will align into a collinear structure of "dumb-bell" type, that is, the electrons are localized on opposite sides of the nucleus at approximately equal distances from it. In this model, the problem reduces to the motion of a particle with mass $m/2$ in a Coulomb field [7.58]:

$$V = -2Ze^2/r - e^2/2r . \tag{7.2}$$

Then the energy of AIS is given by the familiar formula

$$E_n = -2(Z^*)^2/n^2 \mathrm{Ry} , \tag{7.3}$$

where $Z^* = Z - 1/4$, Z is the charge of the nucleus. In the incident-electron energy scale, the resonance position is given as

$$E_{n_{res}} = I_t + E_n , \tag{7.4}$$

where $I_t = E_i - I$; I_t is the ionization energy of the target (for He$^+$, $I_t = 54.5$ eV); E_i is the energy of two-electron ionization (for He, $E_i = 79.01$ eV); I is the energy of a single ionization of the total system (24.58 eV for He). These considerations can be applied not only to the He atom but also to any system

Table 7.3. Energies of AIS states and the positions of corresponding resonances for He, H⁻, He⁻, eV found according to (7.3)

Configuration	E_n	E_{nres}	exp
		He	
$1s^2$	-83.33	—	-79.03[a]
$2s^2$	-20.83	33.67	33.4
$3s^2$	-9.26	45.24	45
$4s^2$	-5.21	49.29	
		H⁻	
$1s^2$	-15.31	—	-14.3[a]
$2s^2$	-3.83	9.77	9.56
$3s^2$	-1.70	11.9	10.22
$4s^2$	-0.96	12.6	
		He⁻	
$1s^1 2s^2$	-3.83	20.75	19.5
$1s3s^2$	-1.70	22.88	22.4
$1s4s^2$	-0.96	23.62	

[a] Stable state

with two valence electrons (ns^2), where Z is the charge of the core. Table 7.3 presents the energies of AIS and the positions of resonances in He, H⁻ and He⁻ ($1sns^2$) calculated according to (7.3). Qualitative agreement with data observed is evident.

Read [7.4, 59] suggested a modified Rydberg formula

$$E_n = -2(Z - \sigma)^2 / (n - \delta_{ns})^2 , \tag{7.5}$$

where σ is an empirical screening parameter ($\sigma \approx 0.25$); δ_{ns} is the quantum defect of the state with an ns^2 core configuration. This formula gives more accurate values of the AIS energies and is applicable to many atomic systems with two electrons outside of the core, i.e., atoms, negative ions, doubly excited states and multiply ionized ions.

Experimentalists have been showing progressively more interest in determining the excitation functions of atoms for excitation from excited states, in particular, metastable states. Optical excitation functions of this sort for He were studied by *Heddle* et al. [7.60]. They determined the functions of the $4^3S - 2^3P$, $4^1P - 2^1P$ and other similar transitions. A resonance structure was also seen in many of these functions.

In analogy with the resonances in the excitation of $1snl$-states of He, the appearance of resonances in the excitation of helium AIS may be predicted too. These resonances are due to the formation of the AIS of the He⁻ ion of $nln'l'n''l''$ type ($n, n', n'' \geq 2$). *Fano* and *Cooper* [7.61] have suggested that, at least, four AIS of this type, $2s^2 2p$ ^2P, $2s2p^2$ ^2D, $2s2p^2$ ^2S and $2p^3$ ^2P, are accessible. Recently six of these AIS have been discovered [7.62]. Processes in which triply excited AIS play a role are beyond the scope of the present book, thus, we refer the reader to the original papers [e.g., Ref. 7.63].

7.3.3 Other Noble Gases

A vast literature is dedicated to the resonances in the electron scattering by other noble gas atoms. It deals, however, mainly with those resonances which arise in the electron excitation from the inner shells or in the excitation of two outer-shell electrons. *Cvejanović* [7.64] have performed experimental measurements of the electron-impact excitation functions for the 6s-states of the Xe atom with a resolution of better than 20 meV, allowing a series of closely lying resonances to be resolved; some of them for the first time. The resonances due to the AIS with $(nsnp^6)mlm'l'$ and $(ns^2np^4)mlm'l'm''l''$ configurations were studied in [7.65–67]. The total and differential excitation cross sections for metastable excitation in Ne, Ar and Kr were measured using the metastable spectroscopy technique [7.8]. The resonances at 16.91 eV $[2p^5(^2P_{1/2,3/2})3s3p\ ^3P]$ and at 18.67 eV $[2p^5(^2P_{1/2,3/2})3p^2\ ^1S,\ ^1D]$ were discovered in electron scattering by the Ne atom. Much useful information on the experimental observation of resonances in low-energy electron scattering by noble-gas atoms can be drawn from the review [7.59]. Theoretical calculations were reported in [7.68] on the resonance scattering by noble-gas atoms. The results of these calculations agree fairly well with the experimental data.

A large number of resonances were found in the excitation functions for metastable states in Ne, Kr and Xe [7.69]. One should note that much fewer works deal with these atoms, while the number of resonances considerably exceeds that observed for lighter atoms. In addition, quite little is known about the classification and coupling schemes appropriate for these resonances [7.70–72].

The electron-impact excitation functions for the $np^5(n+1)$s excited states of Ar and Kr were recently measured in the energy range from 1 to 2.5 eV above the threshold by *Baas* et al. [7.73]. The energy resolution was typically better than 20 meV at scattering angles in the region of 0 to 90°. Resonance features are observed in all the excitation functions studied.

7.4 Helium Ion

The resonance structure in the electron scattering cross sections of the helium ion, which is isoelectronic with the hydrogen atom, is significantly more complex than that for the hydrogen atom because of the presence of a long-range attractive Coulomb potential.

The first experimental study of the excitation cross section of He$^+$ by electrons was carried out in 1966 by *Harrison* et al. [7.74] using the crossed beam method. Although the energy spread of the electron beam was ~ 1.5 eV, a broad maximum was detected in the cross section at an energy of 48 eV.

Daly and *Powell* [7.75] have continued the studies of the excitation of the 2s-level of He$^+$. The excitation efficiency for the 2s-level of the He$^+$ ion was determined by detection of totally ionized He atoms produced in the three-stage process He$(1s^2) + e \rightarrow$ He$^+(1s) + 2e$; He$^+(1s) + e \rightarrow$ He$^+(2s) + e$; He$^+(2s) + e \rightarrow$

He^{2+} + 2e since at electron energies of 40.8 up to 54.4 eV the He^{2+} ions may be produced only via these processes. Since no absolute value of the cross section was obtained in [7.75], only the shape of the measured curve can be compared to that calculated in [7.76]. The theoretical cross section was obtained via CCM and included six states (1s − 2s − 2p − 3s − 3p − 3d) of the He$^+$ ion. The comparison indicates that the origin of the first minimum can be explained by an electron capture into the lowest ^{1}S AIS of the He atom, while the second minimum seems to be caused by a combined effect of electron capture into the ^{1}S and and lowest ^{1}P^0 and ^{1}D states of the He atom.

The most careful experiment on the excitation of the 2s-level of He$^+$ was carried out by *Peart* and *Dolder* [7.77], who used a method analogous to that of [7.74]. The resonance structure that they found agrees well in terms of shape with both the experimental data of [7.74] and the calculated data of [7.76].

An experimental study of the excitation of the 2p-level of He$^+$ was undertaken in 1974 *Zapesochny* et al. [7.78]. They used an apparatus with modulated intersecting electron and ion beams. The efficiency of the exciation was found by measuring the radiation at a wavelength of $\lambda = 304$ Å due to the 1s ^{2}S$_{1/2}$ − 2p ^{2}P$^0_{1/2,3/2}$ transition. These first experiments did not reveal any structure in the excitation function. The first experimental indication of the existence of resonances in this excitation cross section in the 45–55 eV region appeared in 1981. In the recent experiment [7.79] a clearly defined resonance structure was found in the cross section. This structure consisted of broad maxima, two of which lie below the $n = 3$ threshold of He$^+$.

On the theoretical side, the calculations of resonances in the e − He$^+$ scattering cross sections have been carried out by the CCM [e.g., Refs. 7.76, 80–83 and the references therein], by the close-coupling method with correlation functions [7.84, 85], by the algebraic variational method [7.86], and by the DM [7.87]. Let us briefly discuss the application of DM (Chaps. 5, 6) to the excitation of the He$^+$ ion and compare the results of different methods.

We consider the excitation of the He$^+$ ion at energies at which besides the elastic channel, only the 2s and 2p excitation channels are open. Here the resonances caused by AIS converging to the threshold of the $n = 3$ state of He$^+$ must arise.

In order to calculate the wave functions of AIS converging to the $n = 3$ threshold, the following basic configuration sets were used in the CI calculations [7.87]: ^{1}S − 3sns, 3pnp, 3dnd; $3 \leq n \leq 10$.

That choice enables one to describe the typical features of the 2s and 2p-excitation of He$^+$ in detail, with good accuracy. The calculated energies and widths of these AIS, as well as the data of other theoretical calculations are presented in Table 7.4, where, as in Table 5.1, the classification proposed in [7.88] is used. Comparison of the energies and widths of the $(3, n\alpha)^{2S+1}L$ AIS of the helium atom [7.87] with those from other methods shows (Table 7.4) that all data are in good agreement with DM results [7.89] using the diagonalization of interaction matrices of higher dimension, with CCM data for 1s − 2s − 2p states with 20 correlation functions [7.85], and also with data obtained for 1s −

Table 7.4. Energies [eV] and widths [eV] of the $(3, n\alpha)$ ^{2S+1}L AIS of the He atom

Classification	[7.87] a	[7.89] b	[7.85] c	[7.91] d	[7.90] c	[7.88] b	[7.93] [7.92] a
$(3,3a)$ ^1S	69.38	69.38	69.40	69.39	69.39	69.38	–
	$(9.72 - 2)$	$(8.28 - 2)$	$(8.25 - 2)$	$(8.16 - 2)$	$(8.16 - 2)$		
$(3,3b)$	70.47	70.47	70.41	70.38	70.36	70.42	–
	$(1.58 - 1)$	$(1.46 - 1)$	$(2.03 - 1)$	$(1.81 - 1)$	$(1.81 - 1)$		
$(3,4a)$	71.39	71.39	71.39	71.36	71.36	71.36	–
	$(5.64 - 2)$	$(4.70 - 2)$			$(2.04 - 2)$		
$(3,4a)$ ^3S	71.19	71.20	–	–	–	71.19	–
	$(1.85 - 4)$	$(2.04 - 4)$					
$(3,4b)$	71.67	71.67	–	–	–	71.65	–
	$(2.05 - 4)$						
$(3,3a)$ $^1P^0$	69.90	69.90	62.92	69.87	69.89	69.89	69.88
	$(1.83 - 1)$	$(1.51 - 1)$	$(2.04 - 1)$	$(1.90 - 1)$	$(1.89 - 1)$		$(1.50 - 1)$
$(3,4c)$	71.23	71.23	–	–	–	71.23	71.21
	$(5.34 - 4)$	$(5.58 - 4)$					$(5.5 - 4)$
$(3,4b)$	71.46	71.46	–	71.31	71.33	71.44	71.46
	$(6.68 - 2)$	$(6.67 - 2)$		$(3.95 - 2)$	$(2.99 - 2)$		$(6.7 - 2)$
$(3,4a)$	71.67	71.67	–	71.63	–	71.64	71.67
	$(4.96 - 2)$	$(4.77 - 2)$		$(8.44 - 2)$			$(4.6 - 2)$
$(3,3a)$ $^3P^0$	69.46	69.46	69.48	69.47	69.48	69.45	69.44
	$(1.06 - 1)$	$(9.76 - 2)$		$(8.10 - 2)$	$(8.16 - 2)$		$(9.8 - 2)$
$(3,3b)$	70.63	70.64	–	70.59	70.60	70.57	70.61
	$(5.50 - 2)$	$(4.91 - 2)$		$(3.03 - 2)$	$(2.72 - 2)$		$(4.9 - 2)$
$(3,4a)$	71.43	71.43	–	71.40	–	71.40	71.41
	$(2.54 - 2)$	$(2.32 - 2)$		$(3.73 - 2)$			$(2.3 - 2)$
$(3,3a)$ ^1D	69.65	69.65	69.67	–	69.62	69.64	–
	$(1.99 - 1)$	$(2.42 - 1)$	$(1.54 - 1)$		$(1.36 - 1)$		
$(3,3b)$	70.49	70.49	–	–	–	70.45	–
	$(1.28 - 1)$	$(1.15 - 1)$					
$(3,4a)$	71.52	71.53	–	–	–	71.49	–
	$(9.44 - 2)$	$(1.14 - 1)$					
$(3,3a)$ ^3D	70.15	70.15	–	–	–	70.13	–
	$(2.99 - 2)$	$(5.44 - 2)$					
$(3,3b)$	71.31	71.31	–	–	–	71.31	–
	$(1.81 - 3)$	$(1.42 - 3)$					

a – DM; b – CI; c – CCM; d – Complex rotation method

2s – 2p – 3s – 3p – 3d states and three pseudostates [7.90], as well as by the complex coordinate rotation method [7.91]. This indicates that $3ln'l'(n' > 10)$ configurations which are not included give no essential contribution to the wave functions (and energies) of the He atoms AIS in question. We remind the reader that the energies of AIS (Table 7.4) are expressed in a scale where the ground-state He atom energy is set to be zero.

For many fixed principal quantum numbers this method takes into account the angular correlation of electrons quite satisfactoryly, although because the sets of the basic Slater's determinants used are not complete, the radial correlations

Table 7.5. Comparison of theoretical and experimental energies [eV] and widths [eV] of the AIS of the He atom converging to the $n = 3$ state of He$^+$

Classification	Theory	Experiment		
	[7.87]	[7.94]	[7.95]	[7.96]
	a	b	b	b
$(3,3a)\ ^1P^0$	69.90	69.94 ± 0.04	49.919 ± 0.0007	69.917 ± 0.012
	(0.183)		(0.132 ± 0.014)	(0.178 ± 0.012)
$(3,4b)\ ^1P^0$	71.46	–	–	71.30 ± 0.04
	(0.0668)			(0.07)
$(3,4a)\ ^1P^0$	71.67	71.66 ± 0.01	–	71.601 ± 0.018
	(0.0496)			(0.096 ± 0.015)
$(3,5a)\ ^1P^0$	72.21	72.20 ± 0.01	–	72.181 ± 0.015

a – DM; b – photoionization

are taken into account much less accurately. However, the inclusion of the radial correlations in the calculations of AIS is less important than that of angular correlations. It is only important in cases when both electrons are arranged near the nucleus at approximately equal distances from the latter, e.g., as in the case of the $1s^2\ ^1S$-state of the He atom.

Table 7.5 presents the characteristics of the $^1P^0$ AIS of He, where the calculated parameters [7.87] are compared with experimental photoabsorbtion data for the vacuum-UV-region. As can be seen, the calculated data are in good agreement with experimental results. The most remarkable agreement is observed with the results of a recent experiment [7.91] which determined the energy and width of a lower $(3, 3a)$ AIS ($3s3p$ in a single-particle approximation).

Partial width $\Gamma_i(3, n\alpha)$ calculations for the helium AIS show that for most AIS considered the probability of autoionization into a ground $1s$-state of the He$^+$ ion is smaller than 1 % of the total autoionization probability. This agrees well with the results of [7.92] where similar calculations were carried out with simple Coulomb wave functions used in the continuum. Calculations showed that the inclusion of the coupling between the continua to which the AIS may decay strongly influence the fraction of partial widths related to the $n = 2$ states of He$^+$ (due to the energy degeneracy of these levels) [7.87]. At the same time the partial width Γ_1 depends weakly on the coupling between the open channels. This means that the partial widths are extremely sensitive to details of inter-channel interaction in the atomic system.

Thus the calculations of the characteristics of $(2, n\alpha)$ and $(3, n\alpha)^{2S+1}L$ autoionizing states of He have shown that the DM, being more simple and convenient, enables one to obtain the AIS parameters for two-electron systems practically with the same accuracy as the more commonly used CCM.

Figure 7.12 presents some partial excitation cross sections for the $n = 2$ levels of He$^+$ calculated in [7.87], involving 5 lower AIS of He in each partial wave. Note that the shift Δ_μ was not taken into account because of its small value.

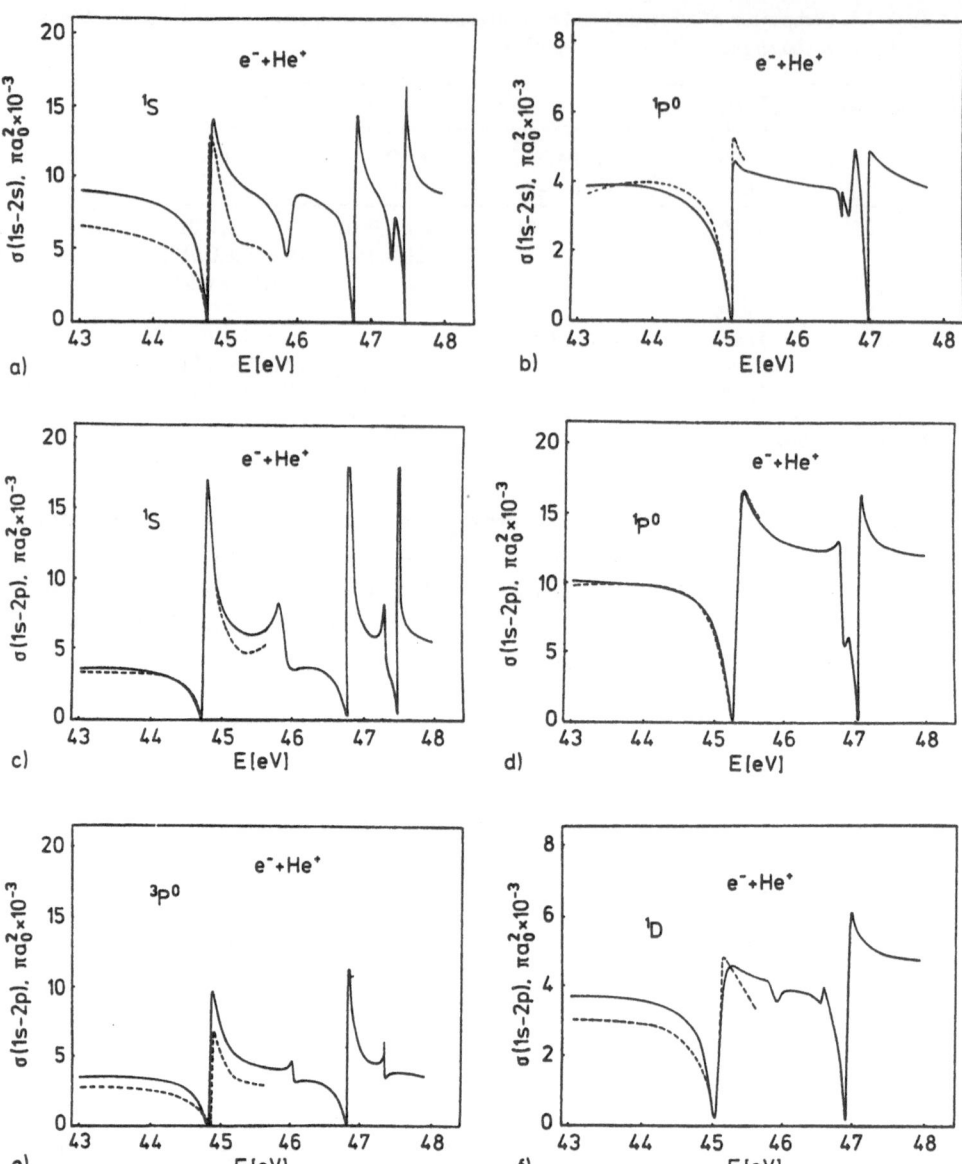

Fig. 7.12a–f. Partial cross sections of the excitation of the 2s and 2p levels of He$^+$. (—) DM calculation [7.87], (- - -) CCM calculation [7.85]

In order to analyze the resonance structure in the cross sections, the latter may be parametrized in the vicinity of an isolated resonance by the following formula from *Fano* [7.97]:

$$\sigma(ij) = \sigma_a(ij)\frac{(\varepsilon + q_{ij})}{1 + \varepsilon^2} + \sigma_b(i,j) \tag{7.6}$$

where σ_a and σ_b are the energy functions which vary slowly in the vicinity of a resonance; $\varepsilon = (E - \varepsilon_\mu)/\frac{1}{2}\Gamma_\mu$; q_{ij} is Fano's profile index which defines the resonance shape. Hence, as is seen from Fig. 7.12, $(3, n\alpha)$ resonances which belong to a principal a-series have two common features. Firstly, for all $(3, n\alpha)$ resonances the profile index $q_{1s,2(sp)}$ has a positive sign which results in the decrease of the excitation cross section in the region below the resonances. Moreover, for the $^1P^0$ and ^1D resonances, the index q also has a very small value which results in pure destructive interference near these resonances. Secondly, for all $(3, n\alpha)$ resonances, σ_b is close to zero which, as shown in [7.97], means that in electron scattering by He$^+$ ions only one out of two linear combinations of 2s and 2p levels of He$^+$ are excited effectively.

Figure 7.12 also presents the results of CCM calculations for the partial cross sections near the lowest $(3, n\alpha)$ resonances [7.85]. As can be seen, DM calculations of the energies, widths and shapes of the resonances are in good agreement with the results of this quite accurate calculation.

In calculating total and differential excitation cross sections in [7.87], the summation was performed up to $L = 4$ in $(4.75, 105, 106)$. The differential cross sections obtained (Fig. 7.13) reveal significant angular asymmetry with pronounced large angle scattering. Thus, the 2p excitation differential cross section at $\theta = 180°$ increases in the lowest $(3,3a)$ ^3D-resonance region more than twice in comparison with that in the nonresonance region. The differential cross section

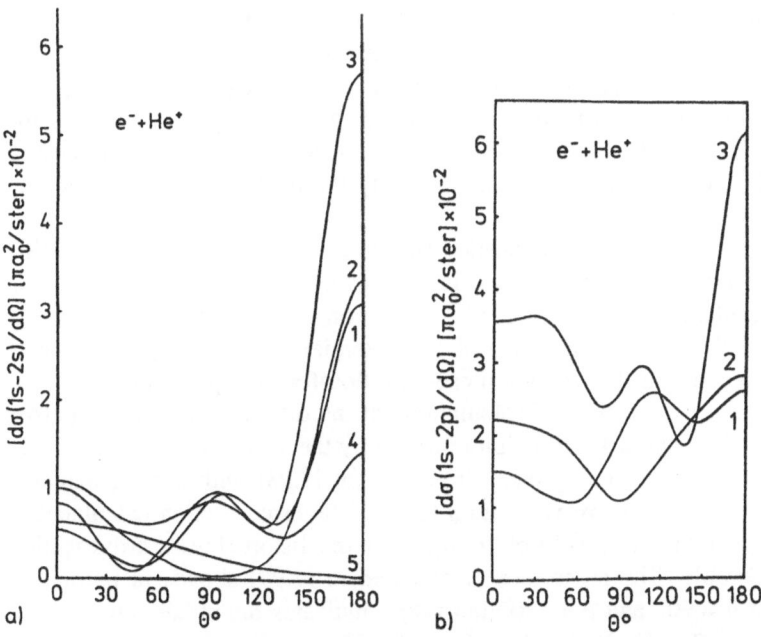

Fig. 7.13. Angular dependence of differential cross sections for electron-impact excitation of the 2s (a) and 2p (b) levels of He$^+$ for various electron energies. $1 - E = 3.1911$ Ry, $2 - E = 3.2921$ Ry, $3 - E = 3.3491$ Ry (calculation [7.87]); $4 - E = 3.20$ Ry (calculation [7.98]); $5 - E = 3.20$ Ry (calculation [7.99])

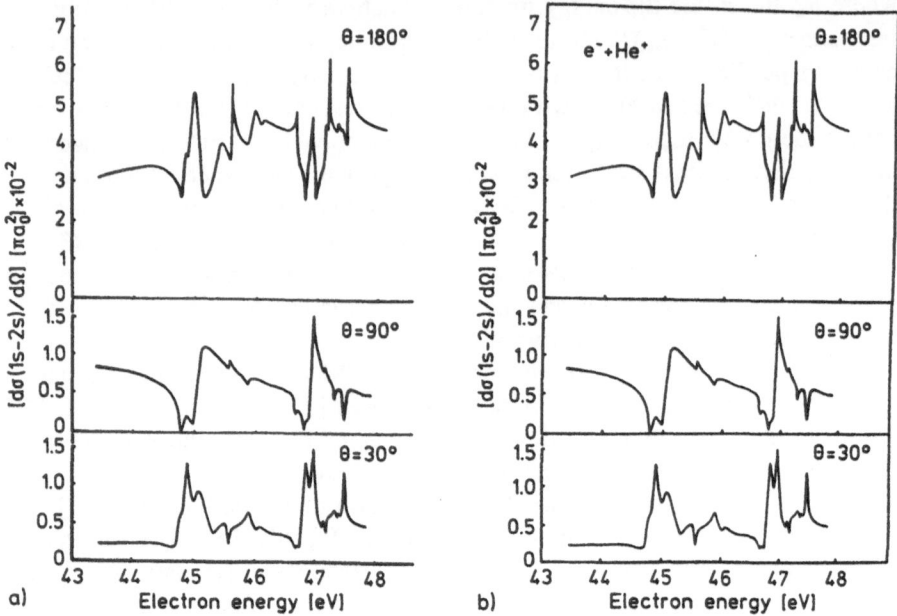

Fig. 7.14. Eneregy dependence of the differential cross sections for electron-impact excitation of the 2s (a) and 2p (b) levels of He+

for 2s exciation for He+ has two minima in the nonresonance region, at $\theta = 50°$ and $\theta = 130°$, and an intermediate maximum at $\theta = 90°$, all due to the interference of S, P and D waves. At the energy of $k_1^2 = 3.2921$ Ry which corresponds to the first minimum in the total cross section [which, in turn, is caused by the lower (3, 3a) ^1S and $^3P_0$ resonances] this structure in the angular dependence of the differential cross section vanishes, and only the deep minimum remains at $\theta = 100°$.

Strong dependence of the resonance structure in the differential excitation cross sections on the scattering angle is also seen in Fig. 7.14. Depending on the scattering angle, the AIS may lead both to constructive ($\theta = 30;\ 180°$) and to destructive ($\theta = 90°$) interference of the potential scattering with resonance scattering, especially in the excitation cross section for the 2p level of He+.

Let us compare the results of calculations of the differential excitation cross section for the 2s-level of the He+ ion obtained by the R-matrix method with 1s – 2s – 2p state coupling being taken into account [7.98] with those calculated within the distorted wave approximation [7.99]. The cross section calculated in [7.87] agrees well with the results of [7.98], while the distorted wave cross section differs considerably. The cross section obtained in [7.99] shows, in particular, no maxima at $\theta = 90°$ and $\theta = 180°$ but rather indicates the maximum value of scattering at $\theta = 0°$ and a minimum at $\theta = 180°$. These differences are apparently related to the fact that in the distorted wave calculation the contribution of the $^{1,3}S$ and $^1P^0$ partial waves was taken into account, whereas the D-waves were not included.

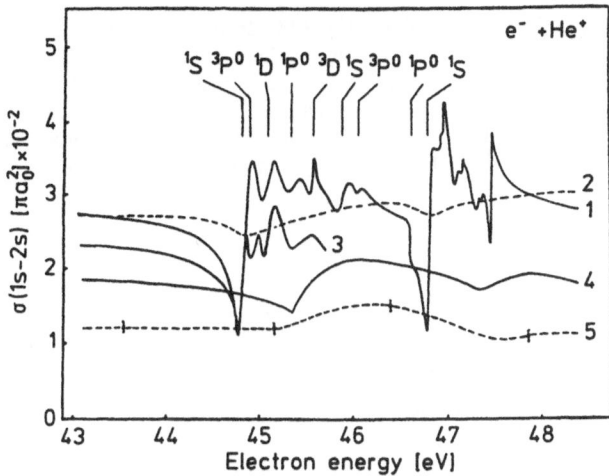

Fig. 7.15. Total cross section for electron impact excitation of the 2s level of He⁺. *1* – found by DM [7.87], *2* – found by DM and convoluted with a Gaussian distribution function of FWHM = 1.5 eV [7.87], *3* – found by CCM [7.85], *4* – experimental [7.77], *5* – experimental [7.75]. The vertical lines mark the positions of the $(3, n\alpha)$ AIS of He

Figure 7.15 shows theoretical and experimental total cross sections for the excitation of the 2s level of He⁺. We see that the results calculated by the CCM with correlation functions and by the DM agree well with one another. Particularly good agreement was achieved for the position and shape of the first deep minimum in the cross section at 44.8 eV; this minimum stems from the low-lying ¹S and ³P⁰ autoionizing states of He. In the nonresonance region, in contrast, the cross section found by DM is 25 % larger than that calculated by the CCM with correlation functions. The reason is that the polarization of the target caused by the higher-lying states is not taken into account as exactly in the DM as it is in the CCM with correlation functions. The theoretical cross sections are significantly larger (by a factor 1.5 – 2) than the experimental cross sections. *Henry's* recent analysis [7.100] of the reasons for such a large discrepancy shows that the experimental cross section reported in [7.77] is apparently substantially too low because of difficulties in the normalization of the experimental curve. At the same time, the shape of the cross section found by DM and averaged over the distribution of electron energies in the beam, 1.5 eV FWHM [according to Ref. 7.77], agrees well with the experimental cross sections. This shows that the electron-electron correlation interaction which is responsible for the resonance effect is taken into account quite accurately in DM calculations.

Figures 7.16, 17 show theoretical and experimental excitation cross sections for the resonance of the 2p level of He⁺. The result of the DM calculations (Fig. 7.16) agree excellently with those obtained by CCM, in contrast to the case of 2s-level excitation, where DM data exceed the CCM data by 25 %. In Fig. 7.17 the DM calculations are convoluted with a Gaussian distribution function of 1.5 eV FWHM, to match the experimental conditions of [7.79]. The shape of

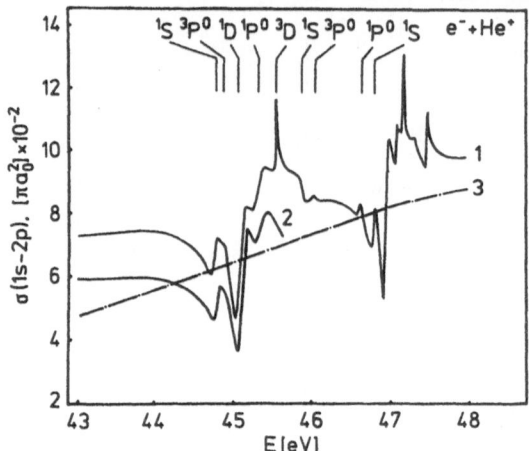

Fig. 7.16. Total cross section for the electron-impact excitation of the 2p level of He$^+$. *1* – found by DM [7.87], *2* – found by CCM [7.85], *3* – experimental [7.78]. The *vertical lines* mark the energies of the $(3, n\alpha)$ AIS of He

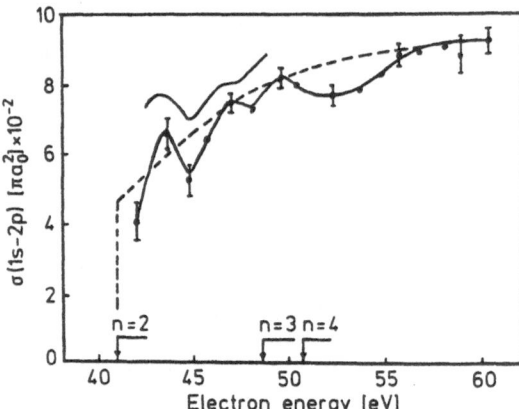

Fig. 7.17. Comparison of the averaged theoretical and experimental cross sections for electron-impact excitation of the 2p level of He$^+$. (●) experiment [7.79], (—) cross section found by DM and convoluted with a Gaussian distribution function with FWHM = 1.5 eV [7.87], (---) experiment [7.78]. The lower solid line is a least-squares fit to the experimental points

the experimental cross section is resproduced quite well, in the energy interval between the $n = 2$ and $n = 3$ thresholds of He$^+$, although the calculated cross section is slightly larger. Consequently, the resonance structure observed in the cross section taken from [7.79] can be explained entirely in terms of an Auger decay of an autoionizing state of He, which converges on the $n = 3$ and $n = 4$ thresholds of He$^+$.

In recent studies [7.101, 102], CCM, with 20 states and pseudostates of He$^+$, was used to calculate the parameters of the autoionizing states of He below the $n = 2$ threshold of He$^+$. It was shown that in the nonresonance energy region the short-range electron-electron correlations are dominant, while in the resonance region, long-range correlations are more important.

In a subsequent paper [7.86], *Oza* carried out the most complete calculation of resonances in e + He$^+$ scattering. For an eleven-state basis set he included 19 channels and for a fourteen-state basis set he used 26 channels. Resonance parameters for 19 F-wave resonances and 10 G-wave resonances were obtained with the outer electron attaining $n = 7$.

In concluding this section, we wish to stress that a detailed comparison of the experimentally and theoretically determined resonance structure in the excitation cross section for the $n = 2$ levels of He^+ will require further experiments in which the energy spread of the electron beam is at most $0.1 - 0.2$ eV. Achieving this figure seems totally realistic in view of the substantial progress which has recently been made in experimental apparatus. Furthermore, it will be necessary to solve the problem of measuring accurate absolute cross sections. Until now "absolute" cross sections have been found by normalizing the experimental curve to the Born-approximation theoretical curve in the region of $100 - 300$ eV. This method is not always justified.

7.5 Alkali Metal Atoms

A distinctive feature of the scattering of slow electrons by alkali atoms is the clearly expressed ^3P shape resonance in the elastic cross section. This resonance was originally predicted in theoretical calculations carried out by CCM of ns–np states by *Karule* [7.103]. Its existence was subsequently confirmed in other theoretical calculations [7.104–108]. Experimentally, the ^3P resonance has been observed in [7.109, 110]. Table 7.6 shows the results of theoretical and experimental studies of the positions of this resonance. The shape resonances in the elastic scattering of electrons by alkali atoms exceed by an order of magnitude the background value of the cross section. This causes changes in the thermal conductivity, the electrical conductivity and the viscosity of a weakly ionized plasma [7.112], because these properties are determined primarily by the interaction of slow electrons with the neutral component of the plasma.

Analyzing the question of the existence of a ^3P shape resonance in $e + Cs$ scattering, *Fabrikant* [7.107] concluded that there is a ^3P bound excited state of the negative Cs^- ion with an electron affinity of ≈ 0.027 eV. However, a final resolution of this question must await calculations using more accurate atomic wave functions which would yield a more accurate value of the Cs atom polarizability.

Like the hydrogen atom, the atoms of the alkali elements can capture an additional electron, forming a stable negative ion in the ^1S state with an electron affinity ranging from 0.61 eV for Li^- to 0.47 eV for Cs^-. All other states of negative ions of alkali elements which have been found to date are unstable; i.e., they are autoionizing states. All these states lie below excited states of the atoms and are seen as resonances in the cross sections for elastic and inelastic scattering of electrons by the neutral atoms. Since the valence electrons in the excited states of the alkali atoms are not degenerated in orbital angular momentum l, in contrast to the hydrogen atom, only a limited number of autoionizing states can lie under each threshold. This is, in fact, a consequence of the short-range character of the potential [7.113].

7.5.1 Lithium Atom

Burke and *Taylor* [7.108] pointed out the existence of ^{1}D and ^{1}P AIS of the Li$^-$ ion below the excitation threshold of the 2p ^{2}P^0 state of Li. These AIS were not found in multi-configuration Hartree-Fock calculations [7.114]. However, these calculations did reveal four AIS below the excitation threshold of the 3p ^{2}P^0 state and three AIS below the 3s threshold. Later, the same AIS were analyzed using the CI method [7.115, 116]. The energy of the ^{1}P^0 and ^{1}D AIS below the threshold of the 2p state was found to be 1.85 eV, i.e., somewhat larger than the threshold energy [7.115]. This is related, evidently, to the fact that a restricted wave function basis was used in the CI calculation. The energies and the widths of the ^{1}S, ^{3}P^0 and ^{3}S AIS which lie below the 3s threshold were calculated by DM [7.115]. The energies of the ^{1}D and ^{1}S states below the 3p threshold were obtained by the CI method. The obtained energies differ from the results of [7.114] by not more than 0.06 eV (3.15, 3.29, 3.36 and 3.74 and 3.75 eV, respectively.). Figure 7.18 shows the energy diagram of the Li$^-$ ion AIS as calculated in [7.108–114, 117].

The scattering of electrons by Li atoms has been the subject of several experimental [7.109, 110, 118, 119] and theoretical studies [7.103, 105, 108, 115, 120]. However, the amount of the information derived is far smaller than for other alkali metals. In particular, prior to the appearance of [7.115] there had been no theoretical study of the formation of AIS in experiments on scattering, and this theoretical work was limited to resonances in the elastic channel.

Below we shall present the results of DM calculations for elastic scattering of electrons by Li and Na atoms as well as excitation of resonance levels, which take into account the possibility of temporary AIS formation with subsequent decay into the elastic channel (ns-level) and into the excitation channel (np-level) [7.115, 116].

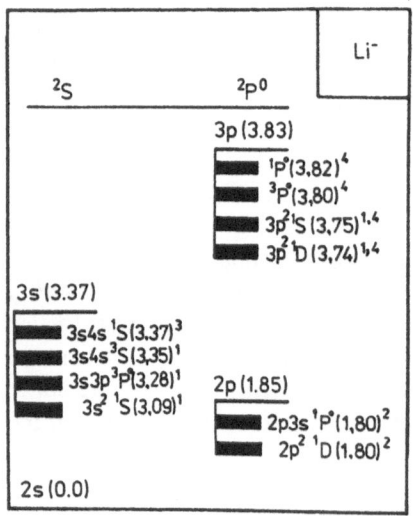

Fig. 7.18. Energies of AIS of the Li$^-$ ion. (—) Li levels, (▬) Li$^-$ levels. The energies, in eV, are given in parentheses

Since the alkali atoms have only a single electron outside the filled shells, it is quite resonable, in the first approximation, to consider the transitions of the optical electron in the field of a frozen core in order to take into account the response of the alkali target to external perturbations. However, when studying the interaction of an optical electron with the atomic core one must take into account the polarization of the core, which considerably influences the oscillator strength distribution. Moreover, one must include the polarization of the atom due to the interaction with the incident electron. Since the coupling of the ground state of the alkali atom with the first resonance state is responsible for the main part of the polarizability, the elastic scattering in the nonresonance region may be described quite accurately by taking into account the coupling of only these states.

The system of equations for the case of electron scattering by complex targets was written in its general form in Chap. 5. Its solution [7.155, 116] was obtained using the *IMPACT* routine [7.121].

For the Li atom, the Hartree-fock wave functions were chosen as the target wave functions of the AIS, whereas in the case of the Na atom the phenomenological polarization potential of the core was added to this Hartree-Fock potential (see next section for more details). For Li only the exchange between the incident electron and the valence electron was taken into account, while for Na exchange with both the valence electron and with core electrons was taken into account.

Both Feshbach resonances and shape resonances with threshold features arise in electron collisions with alkali metal atoms. In general, the cross section of electron scattering by an Li atom in the near-zero energy region is dominated by the $^3P^0$ resonance [7.103, 108].

Feshbach resonaces are clearly revealed in the partial cross sections (Fig. 7.19) but due to the relatively weak contribution of these partial waves to the total cross section in a given energy region they are manifested only weakly in the total

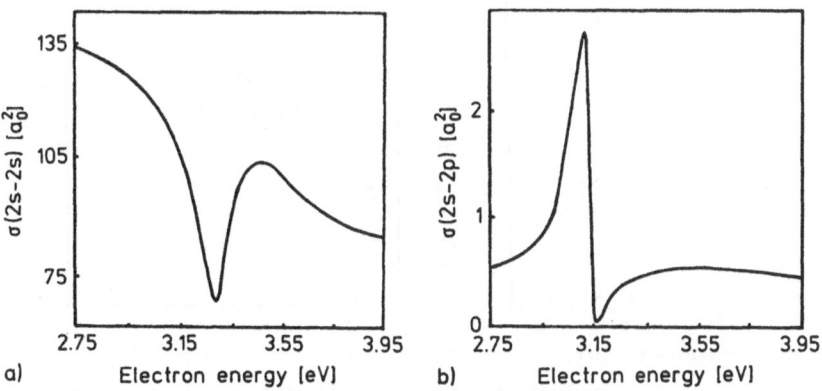

Fig. 7.19a,b. Partial cross sections for e + Li scattering [7.7]. (a) $^3P^0$ cross section of elastic scattering; (b) ^1S cross section of excitation of the 2p level of Li

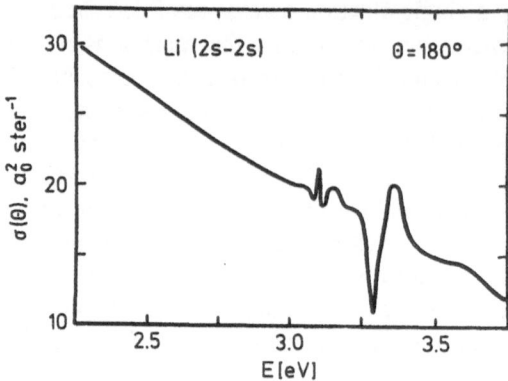

Fig. 7.20. Differential cross section for backward elastic scattering of electrons by the Li atom [7.7]

elastic and total excitation cross sections for the 2p level. Figure 7.20 shows the differential cross sections for large angle e + Li scattering as a function of energy, which show more details of the scattering processes. In the differential cross sections the resonances are pronounced fairly well, in contrast to the case in the total cross sections.

Both experimental and theoretical studies [7.115, 122] were carried out on the differential cross sections for the elastic and inelastic scattering of electrons by the Li atom in the energy region above the 2p $^2P^0$ level through an angle of 90°. The energy spread of the electron beam was 0.3 eV. The calculations were carried out by DM including two open channels. Resonances were observed quite clearly in the calculated partial cross sections. The contributions of resonances to the total and differential cross sections were less pronounced. The reason was that the resonances were manifested in only the ^1S, ^3S and $^3P^0$ waves, which correspond to partial cross sections which are small in comparison with those for the D, F and G waves which dominate the cross section. Figure 7.21 shows

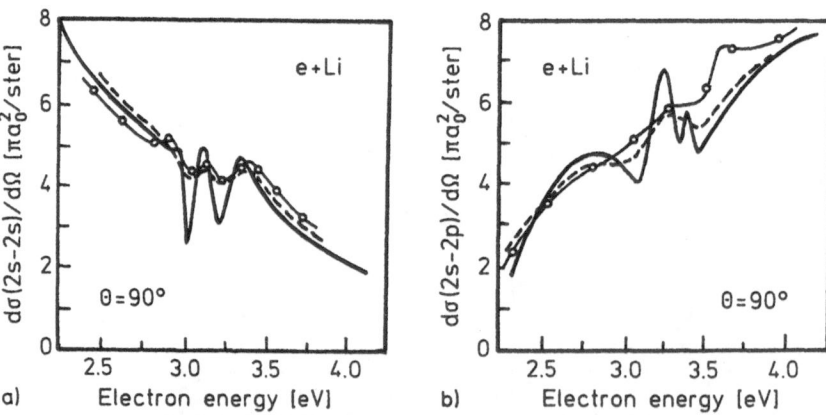

Fig. 7.21. Differential cross sections at 90° for (a) elastic e + Li scattering and (b) excitation of the 2p level as a function of energy. (——) calculated by DM, (---) calculated by DM and convoluted with a Gaussian distribution function of FWHM = 0.3 eV, (–o–) experimental. From [7.115]

the results of [7.115]. Since the magnitude of the differential cross section was measured in arbitrary units, to compare it to calculations the experimental curve was shifted vertically to minimize the discrepancy between the two curves in a region remote from resonances. The contribution of the ^{1}S state at 3.09 eV and the combined contribution of the ^{3}P, ^{3}S and ^{1}S states between 3.28 and 3.37 eV are clearly shown in this figure. The resonance contribution to the cross section below the 3d threshold was not taken into account in the calculations, thus the resonances are not seen in Fig. 7.21, which shows only the part of the curve which was measured in [7.115]. The structure of the cross section in this figure is due to the resonances below the 3s threshold. The ^{3}P^0 autoionizing states make the largest contribution to the resonance part. These contributions are especially pronounced in the partial cross section (Figs. 7.19, 20).

7.5.2 Sodium Atom

Autoionizing states of Na$^-$ have been studied by the CCM [7.104, 105] and by the CI method [7.114, 116]. In [7.104, 114], the ^{1}P, ^{1}D, and ^{1}S autoionizing states with energies of 2.0 eV for ^{1}P, ^{1}D and 3.04 eV for ^{1}S states were predicted (the energies were referenced to the ground state of the Na atom).

Figure 7.22 shows the results of more comprehensive calculations [7.116]. In that study, up to thirty Hartree-Fock configurations calculated with the inclusion of the polarization of the $2p^6$ ^{1}S core were chosen as the basis. The polarization of

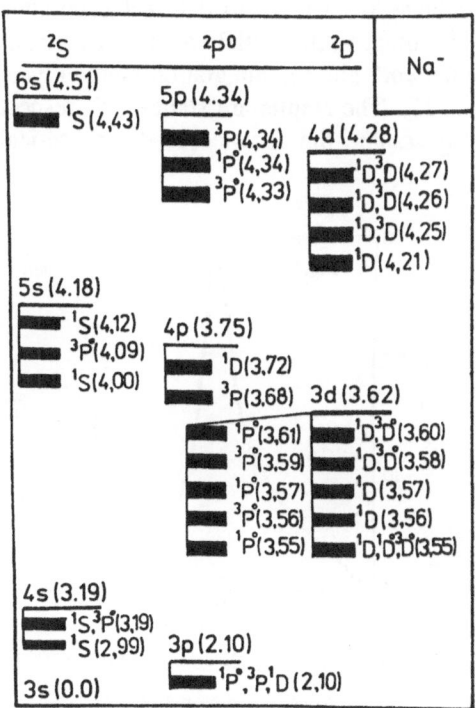

Fig. 7.22. Energies of AIS of the Na$^-$ ion. (—) Na levels, (▬) Na$^-$ levels. The energies, in eV, are given in parentheses

the core was taken into account by introducing a phenomenological potential in the Hamiltonian. The calculations revealed 28 AIS lying below the 6s threshold of the Na atom. Calculations were not done above this threshold. Among these AIS there are also three low-lying states, ^1P, ^1D, and ^1S, which were predicted in [7.123, 124]. As we see in Fig. 7.22, several AIS with an identical term form below the 3d and 4d thresholds. This result demonstrates that the Na atom has the strongest attractive potential in the excited ^2D states.

As calculations show [7.116], the Na$^-$ ion, in contrast to the Li ion [7.114, 115] has a single Feshbach resonance below the 4s threshold which lies quite far from the threshold, and the $^3P^0$ resonance lies just at the threshold (within the accuracy of the calculation).

The existence of the $^1P^0$ and ^1D Feshbach resonances below the 3p level of the Na atom has been confirmed by the diagonalization of the closed channel subspace. Earlier calculations [7.104, 120] gave resonance energies of $E \approx 2$ eV, while the calculation [7.116] gives $E = 2.10$ eV, i.e., again the threshold energy. In order to determine more accurately the energies of the resonances and to find their possible energy shift, the behaviour of the $^1P^0$ and ^1D partial cross sections in the near-threshold region has been studied in detail within the CCM with two channels taken into account [7.116]. The results of these calculations are presented in Figs. 7.23, 24. Comparing the results of these two methods we concluded that the peaks in the $^1P^0$ and ^1D waves at 2.10 eV are caused by the combined influence of the Feshbach resonance and the effect of the opening of a new channel. The partial cross section behaviour in the near-zero energy region is quite curious. In order to eliminate, or at least, to reduce the possible numerical errors of the integration of the integro-differential equation by means of the *IMPACT* routine, the integration network and the integration method were varied in the asymptotic region [7.116, 125]. The results are stable with respect to these variations which allows one to assume the absence of the numerical irregularities.

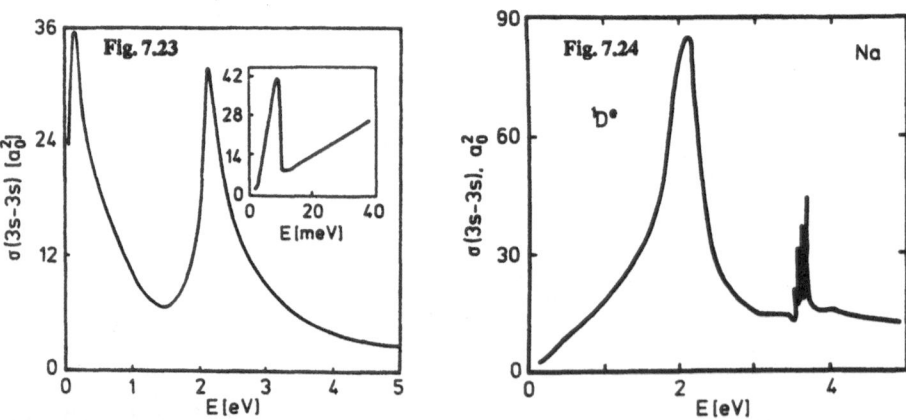

Fig. 7.23. Partial $^1P^0$ cross section for elastic scattering of electrons by Na atoms [7.125]

Fig. 7.24. Partial ^1D cross section for elastic scattering of electrons by Na atoms [7.125]

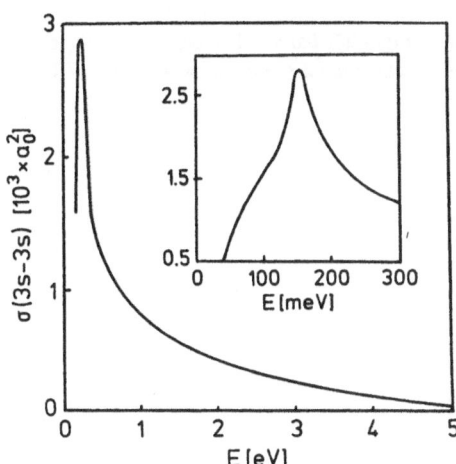

Fig. 7.25. Partial $^3P^0$ cross section for elastic scattering of electrons by Na atoms [7.125]

Table 7.6. Energies [eV] of the ^3P shape resonances in elastic scattering of electrons by alkali metals

Elements	Theory				Experiment	
	[7.104] a	[7.105] b	[7.111] a	[7.106] a	[7.109]	[7.110]
Li	–	0.06	–	–	–	–
Na	0.13	0.10	–	–	0.08 ± 0.02	–
K	–	0.0024	0.02	–	–	–
Rb	–	–	–	–	0.05	–
Cs	–	–	–	0.00075	–	0.15 – 0.30

a – CCM; b – Matrix variational method

As in e + Li scattering in electron scattering by sodium atoms, the ^3P shape resonance is strongly pronounced in the near-zero energy region [7.115] (Fig. 7.25). As is seen, the calculated position of the resonance coincides with the experimental data shown in Table 7.6. However, additional features are present in the ^3S, ^1P, and ^1D partial waves at lower energies. The feature in the ^3S partial cross section is due to the famous Ramsauer effect. The features in the ^1P and ^1D waves has a similar origin, it is caused by the presence of a long-range polarization potential. Indeed, as one may easily obtain by the argumentation presented in [7.126], $\sigma(l = 1) \approx (12\pi/k) \sin^2[k^2(ak + \pi\alpha/15)]$ at low k. This cross section at $a \approx 0$ has the form shown in Fig. 7.23.

Figure 7.26 shows the partial cross section ^1S for e + Na scattering. In the case of elastic ^1S scattering the possibility of a temporary quasistationary state to be formed (upon inclusion of closed channels) has an influence only in a narrow energy interval. For the excitation process, however, inclusion of closed channels does affect the cross section over a wider energy range. As yet we do not have adequate identification of the peak in the partial ^1S cross section when the energy tends to zero. It is probably associated with the presence of the stable Na$^-$ ion.

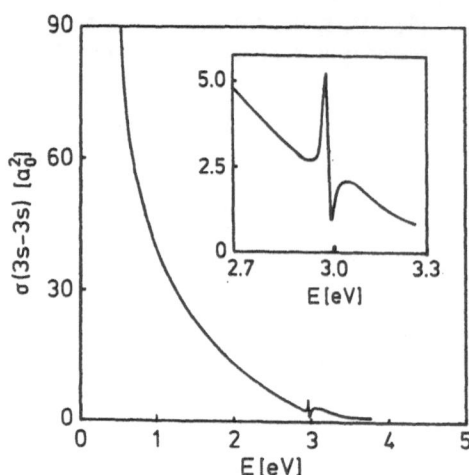

Fig. 7.26. Partial ^{1}S cross section for elastic scattering of electrons by Na atoms [7.125]

The Na$^-$ ion has a series of ^{1}D Feshbach resonances which converge to the 3d and 4d thresholds. These resonances in the elastic ^{1}D wave are quite intense but not as high as the peak at the 3p threshold. Constructive interference dominates the excitation cross section for the ^{1}D resonances converging to the 3d threshold. Feshbach resonances are revealed in the form of narrow bell-like peaks, and destructive interference is typical for the ^{1}D resonance below the 4p threshold.

The diagonalization of the subspace of closed channels carried out in [7.116, 125] shows the existence of a series of Feshbach resonances above the higher thresholds. They must affect the excitation of higher levels more significantly and, especially, the scattering by excited atoms. The reason is that the S, P, and D partial waves with Feshbach resonances are dominant in these processes.

Note that the even P AIS and the odd D AIS do not contribute to elastic scattering because of the conservation of parity and total orbital angular momentum (within the LS-coupling approximation). These states should be observed in photoabsorption, superelastic scattering and in excitation from the 3p level. The characteristic feature of alkali atoms is the existence of AIS which are extremely close to the thresholds from below. Due to this fact they must influence the excitation cross sections in such a way that the latter must be finite at threshold.

One of the earliest experiments on the scattering of electrons by the Na atom were those presented in [7.127, 128], which measured the total cross section. In those experiments, however, an overly large energy step was chosen, and no resonant features of any sort were observed. *Kazakov* et al. [7.129] have recently carried out a more accurate experiment, with an electron beam with an energy spread of 150 meV and with a small energy step. They determined the differential cross section for elastic electron scattering through an angle of 90°, and found many features. These results are compared in Fig. 7.27 with those calculated within the DM. It can be seen from this figure that there are clearly expressed maxima in the cross section which corresponds to the ^{1,3}P^0, and ^{1}D AIS which lie below the 4s, 3d, and 4p thresholds.

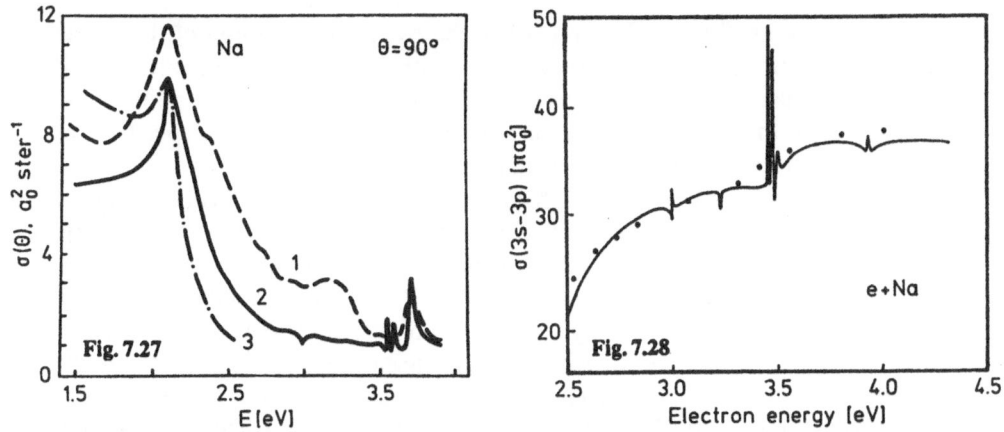

Fig. 7.27. Differential cross section for elastic e + Na scattering. *1* – experimental [7.129], *2* – calculated by DM [7.130], *3* – experimental [1.28]; $\theta = 90°$

Fig. 7.28. Total cross section for electron-impact excitation of the 3p $^2P^0$ level of the Na atom. (—) calculated by DM [7.7], (•) experimental [7.3]

The widths of the series of ^{1}D resonance below the 3d and 4p thresholds have been calculated in [7.130]. These resonances appear to be rather narrow. So, the ^{1}D resonances at 3.55 eV, 3.56 eV, 3.57 eV, 3.58 eV, and 3.60 eV have widths of 3.0 meV, 0.1 meV, 5.4 meV, 0.7 meV, and 5.0 meV, respectively. As our calculation experience shows, the resonance widths are quite sensitive to the choice of the wave function of the continuous spectrum: if an orthogonality condition for the discrete spectrum functions has not been applied to this function, then the widths may increase by an order of magnitude. Note that the influence of the ^{1}S and $^3P^0$ AIS below the 4s threshold at all scattering angles is considerably smaller than that of the ^{1}D states. Near the excitation threshold for the 3p $^2P^0$ level of the Na atom, a feature has been observed which stems from the effect of the $^1P^0$ and ^{1}D AIS of the Na$^-$ ion and from the opening up of a new channel in the S-wave scattering. We have carried out (together with I. Cherlenyak) a theoretical study of the influence of the higher AIS on the total elastic and inelastic cross sections for e + Na scattering. The results are shown in Fig. 7.28. A comparison of Figs. 7.27 and 7.28 clearly indicates that a study of resonances through a measurement of differential cross sections has clear advantages over measurements of total cross sections. At high incident electron energies a considerable number of structural singularities is observed with the energy positions which agree well with those for autoionizing states (Fig. 7.27).

7.5.3 Potassium Atom

Moores [7.111] used CCM in the approximation of 4s-4p-3d states to calculate the phase shifts in the elastic scattering of electrons by the K atom. It was found that the $^1P^0$ and $^3P^0$ phase shifts have resonance behaviour, which stems from the formation of AIS of the K⁻ ion below the excitation threshold of the 4p ^{2}P state. Green's attempt [7.131] to confirm the existence of these AIS in calculations using HSC (hyper-spherical coordinate) was unsuccessful. *Lin* [7.117] pointed out that the reason for this was the use of a very crude approximation, specifically, the use of the Herman-Skillman potential.

Experiments reveal resonant behaviour of the differential scattering cross section between the thresholds of the excitation of the 4p and 5p levels [7.132]. Such behaviour was attributed to the formation of AIS of the K⁻ ion with total angular momenta $L = 0$, $L = 1$ (or $L = 3$), and $L = 2$ with energies of 2.4; 2.68, and 2.6 eV, respectively. However, no theoretical description of these structural features in the cross section has yet been proposed. Detailed data on the differential cross section for elastic scattering and excitation of the 4p level were reported in [7.129]. These data are presented in Fig. 7.29. The AIS energies for the K⁻ ion have been calculated by the CI method and are shown in Fig. 7.30 [7.116]. The positions of most calculated AIS correspond to features observed in Fig. 7.29. However, some features in Fig. 7.29 are not caused by those AIS which are shown in Fig. 7.30. (Calculations of AIS of Li⁻, Na⁻, K⁻ have also been reported in [7.113, 133, 134].)

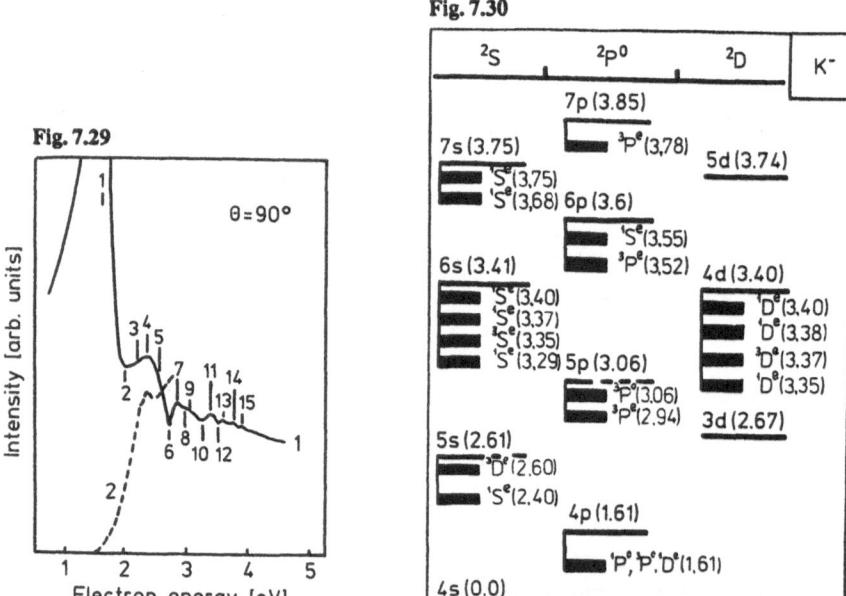

Fig. 7.30

Fig. 7.29. Energy dependence of the differential cross section of e + K scattering [7.129]. 1 (*solid line*) – elastic scattering, 2 (*dashed line*) – excitation of the 4p level; $\theta = 90°$. *1,2,3* ... indicate the numbers of observed features

Fig. 7.30. Energies of AIS of the K⁻ ion. (—) K levels, (▬) K⁻ levels. The energies, in eV, are given in parentheses

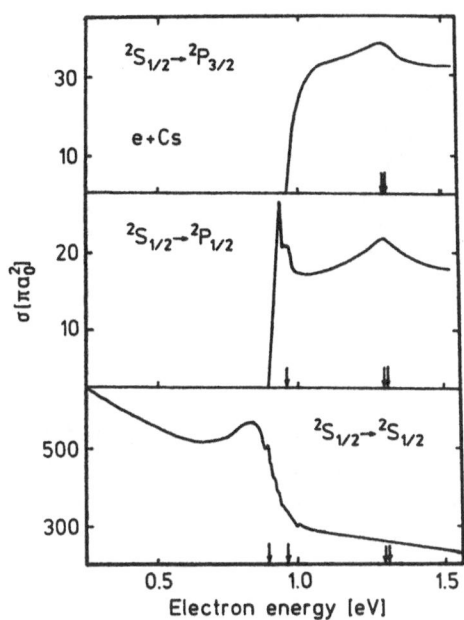

Fig. 7.31. Total elastic scattering cross section (*bottom*) and cross sections for electron-impact excitation of $^2P_{1/2,3/2}$ levels of Cs [7.138]. The *arrows* show the $^2P_{1/2}$, $^2P_{3/2}$, $^2D_{1/2}$ and $^2D_{3/2}$ thresholds

7.5.4 Caesium Atom

The scattering of low-energy electrons by Cs atoms has been the subject of many studies [7.134–138]. As of yet, however, we have no results of a detailed experimental study of resonances although such experiments were planned [7.138]. Our primary source of information of the AIS of the Cs$^-$ ion is thus the theoretical works. The most comprehensive calculations on resonances in e + Cs scattering were carried out by the R-matrix method with a spin-orbit interaction [7.137, 138]. It was found that the cross sections have a complex structure near the 5p $^2P_{1/2,3/2}$ and 5d $^2D_{3/2,5/2}$ excitation thresholds. Figure 7.31 shows the total cross sections for elastic scattering and for excitation of the $^2P_{1/2,3/2}$ levels.

To find and classify resonances, *Scott* et al. [7.138] analyzed the energy dependence of the sum of eigenphase phase shifts (i.e., the phase shifts of the matrix elements found as a result of the diagonalization of the S-matrix) for each fixed $J = 0, 1, 2, 3$ and 4, and for a fixed parity $\pi = \pm 1$. They found 25 resonances; to all of which except two they assigned a definite configuration (Table 7.7).

In the table each family of resonances which have the same configuration, angular momentum, and spin and which lie below the same threshold are denoted by a capital letter. Among the resonances there are some (A and K in Table 7.7) which arise from the capture of the incident electron at a large distance by the long-range potential of the excited Cs atom. We should emphasize, however, that the classification of resonances on the basis of the definite configurations assigned to them is extremely approximate because of the strong inter-configurational interaction. With regard to the positions of these resonance, we note that the A,

Table 7.7. Configurations of resonances observed in e + Cs scattering [7.138]

Designation	Configuration	Designation	Configuration
A	$6pns\,^3P_{0,1,2}$	G	$6p5d\,^3D_{1,2,3}$
B	$6p5d\,^3D_{1,2,3}$	H	$6p^2\,^1S_0$
C	$6s6p\,^1P_1$	I	$6p^2\,^3P_{0,1,2}$
D	$6p5d\,^3F_{2,3,4}$	J	$6s5d\,^3D_{1,2,3}$
E	$6p5d\,^1P_2$	K	$6pnp\,^3D_{1,2,3}$
F	$6p5d\,^1F_3$		

D, H, and I resonances lie directly below the $^2P_{1/2}$ threshold; C, E and K lie between the $^2P_{1/2}$ and $^2P_{3/2}$ thresholds; B and F lie between the $^2P_{3/2}$ and $^2D_{3/2}$ thresholds, and G lies between the $^2D_{3/2}$ and $^2D_{5/2}$ thresholds. The resonances are less obvious in the total cross sections because many resonances of different symmetries overlap and interfere strongly.

7.5.5 Comments on Alkali AIS

We would like to offer a few generalizing comments regarding the role played by electron-electron correlations in the formation of autoionizing states of negative ions of alkali atoms. The model of independent electrons is applicable in most cases for describing singly excited states of atoms. In this model, it is assumed that each electron is moving in the field of the nucleus and of the other electrons and has its own wave function. Each state of the atom is characterized by a definite configuration. This model, however, gives a poor description of the atomic AIS. It has proved to be a particularly unreasonable choice in attempts to describe the negative ion AIS, since electron-electron correlations play an even greater role in this case.

All calculations which have been carried out on the autoionizing states of the ions of alkali atoms indicate strong configurational mixing. The nature of this mixing is such that one can (similarly to the He$^-$ ion) distinguish two limiting extremes for AIS formation: the intra-shell and inter-shell-type [7.69]. In the first case, the excited atomic electron and the captured electron form a highly correlated pair and move at approximately the same distance from the nucleus. Here, the radial correlation in the motion of electrons is important. In the second case, the incident electron is captured at a great distance. The average distances from the electrons to the nucleus are different and there is a pronounced angular correlation in their motion. Correlations in the motion of two excited electrons were analyzed in detail in an excellent review by *Fano* [7.139].

The most popular methods for calculating the parameters of the AIS of negative ions are the Hartree-Fock multiconfiguration method and the CI method. These methods, however, do not generate anything in the way of a natural classifiaction scheme for the AIS. For this reason, considerable interest has been attracted by the HSC-method (Chap. 6). This method, however, is rather complicated and therefore has been used in only a few calculations besides those for He.

At present we have far more theoretical than experimental information on resonances in the collisions of electrons with alkali atoms. As we have seen, in the case of light alkali atoms the resonances lie nearly at the thresholds and very close to each other. High-resolution apparatus is required to observe them. Since the resonances are seen far more clearly in the differential cross sections than in the total cross sections for Li, Na and K atoms, we need more experimental data on the differential cross sections, especially on their absolute values in order to observe and reliably identify these resonances. It would also be extremely valuable to have results obtained from experiments with polarized (Sect. 7.8) monoenergetic electron beams and alkali atoms. On the theoretical side, we need more accurate calculations of the parameters of the resonances, incorporating electron-electron correlations at higher accuracy. We also need new methods for calculating cross sections for stepwise excitation.

Atomic copper since it also has half-filled ns valence shell might behave similarly to the alkali atoms. The first ab initio close-coupling calculations of the electron-impact excitation of atomic copper in the 0.1 – 0.8 eV energy range has been carried out by *Henry* et al. [7.140]. The calculated cross section reveal a rich resonance structure due to Cu^- states which have not been previously known.

7.6 Alkaline-Earth and Alkali Ions

The reason for the interest in ions of alkaline-earth elements is that they play an exceedingly important role in astrophysical and fusion process, in experiments in the space environment of the Earth, in laser technology, and in other promising scientific and technological directions. For example, in the early 1950s astrophysical studies of the solar chromosphere established that the radiation at a wavelength of $\lambda = 2800$ Å, which corresponds to the 3s $^2S_{1/2}$ – 3p $^2P_{1/2,3/2}$ Mg^+ transition, is the second most intense feature after the L_α lines of atomic hydrogen [7.14]. The radiation from ionized calcium and barium is also important in astrophysics. In particular, resonance K and H lines of Ca^+ have yielded much information about the solar chromosphere.

A promising direction in the development of new pulsed gas lasers is to make use of the lasing action of self-terminating transitions of Ca^+ and Ba^+ ions and also of the afterglow of discharges involving transitions of Mg^+, Ca^+, Sr^+ and Ba^+ [7.142, 143].

Resonances can be expected to be far more numerous in the scattering of electrons by ions than in electron scattering by atoms. The reason is that, as was shown by *Presnyakov* et al. [7.144], there is an infinite series of resonances below each excitation threshold of the ion, in contrast with the situation in atoms. Furthermore, the number of resonances should also increase with increasing charge

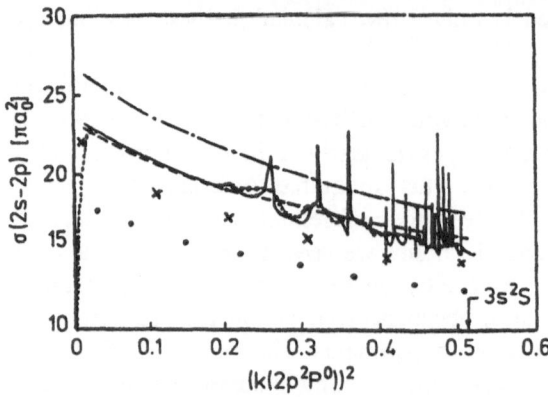

Fig. 7.32. Total cross section for electron-impact excitation of the 2p level of Be⁺. (—) DM calculation [7.152, 156], (···) DM convoluted over a Gaussian distribution function with FWHM = 0.3 eV [7.152, 156], (- - -) CCM calculation including 2s and 2p states [7.157], (- · -) found in the Coulomb-Born approximation [7.157], (●) experimental [7.147], (x) CCM calculation including 2s, 2p, 3s, 3p, 3d states [7.168]

of the ion [7.145]. New experiments have recently yielded the first observation of a clearly defined resonance by low-energy electrons [7.146].

Experimental data on cross sections for the elastic electron scattering by alkaline-earth ions are not yet available. Excitation cross sections have been measured, however, although the resonance structure is still poorly studied.

For the Be⁺ ion, only a single experiment has been done, by *Dunn* et al. at JILA [7.147]. They found the excitation cross section for the 2p $^2P^0$ resonance level; however, no resonance structure was observed near threshold (Fig. 7.32).

The first experiment on the excitation of the 3p $^2P^0$ resonance level of the Mg⁺ ion was reported in [7.148]. The excitation cross section was measured at energies ranging from the excitation threshold up to 100 eV. Significant structure was observed near the threshold; *Kel'man* et al. [7.148] originally attributed this structure to a cascade filling of an Mg⁺ resonance level. Later, the same investigators used an optical method in more careful experiments [7.146], with an electron beam with a much smaller energy spread [0.3 eV against ∼ 1 eV in Ref. 7.148]. As a result, a very clearly defined resonance peak was found in the excitation cross section; this structure was attributed to electron capture into AIS of Mg (Fig. 7.33).

The first experiment on electron-impact exciation of the resonance lines of the Ca⁺ ion was done by *Dunn* and *Taylor* [7.149], who pointed out a dip in intensity near 5 eV. They stated that this feature, lying 1.5 eV below the 5s level, is probably associated with the AIS of the 5snl and 4dnl series. However, the poor energy resolution (about 0.3 eV) did not allow them to come to more clear conclusions.

More recently the same group of Uzhgorod experimentalists has investigated resonances in the cross sections for the excitation of the np $^2P^0$ levels of the Ca⁺(n = 4), Sr⁺(n = 5) and Ba⁺(n = 6) ions [7.7, 150, 151] with an 0.2 eV

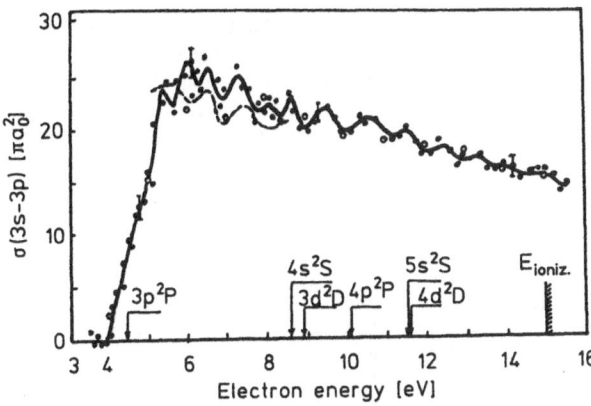

Fig. 7.33. Total cross section for electron-impact excitation of the 3p level of Mg$^+$. (---) DM, convoluted with a Gaussian distribution function with FWHM = 0.3 eV [7.146], (•) experimental [7.146], (—) analysis of the experimental results by a digital filtering method, (o) experimental [7.148]

energy resolution. Since the fine splitting of the ^{2}P levels in the ions of heavy alkaline-earth elements is significant ($\sim$ 0.028 eV for Ca$^+$, $\sim$ 0.099 eV for Sr$^+$ and 0.210 eV for Ba$^+$), it turned out to be possible to study separately the resonance structure in the exciation cross sections of the substates with $j = 1/2$ and $j = 3/2$.

Turning to the theoretical calculations of resonances in the electron scattering by alkaline-earth ions, we note that so far such calculations have been carried out only for the differential cross sections of elastic scattering by Be$^+$ and Mg$^+$, via DM [7.152b] and the excitation cross sectionss of the 2p ^{2}P^0 level of Be$^+$ via DM, the 3p ^{2}P^0 level of Mg$^+$ and the 4p level of Ca$^+$ by the DM Coulomb-Born and CCM [7.146, 152–159]. No calculations of any sort have been performed on the resonances in excitation cross sections of the Sr$^+$ and Ba$^+$ ions. In all the experimental total cross sections a maximum is observed that is not reproduced in theoretical calculations. In particular, the calculations of *Mitroy* et al. [7.154] not confirm the occurrence of a dominant maximum in the excitation threshold for the Ca$^+$ ion (Fig. 7.34). This enhancement is not due to the influence of the AIS which are usually farther from the threshold but is caused perhaps by a sharp increase of the direct excitation cross section at threshold and subsequent fall-off with increasing energy. This effect is less pronounced in the exciation cross section for the Mg$^+$ ion (Fig. 7.33) but is manifested clearly for the Sr$^+$ and Ba$^+$ ions (Figs. 7.35, 36).

7.6.1 AIS of Beryllium and Magnesium Atoms

Consider first the 2pnl AIS of Be. We shall use the criterion for choosing the states which has been applied in CCM. In order to obtain accurate characteristics of the resonances related to some $nl\varepsilon l$ channel, it is necessary to include the higher $n'l'$ states of the target into the CCM expansion which gives the maxi-

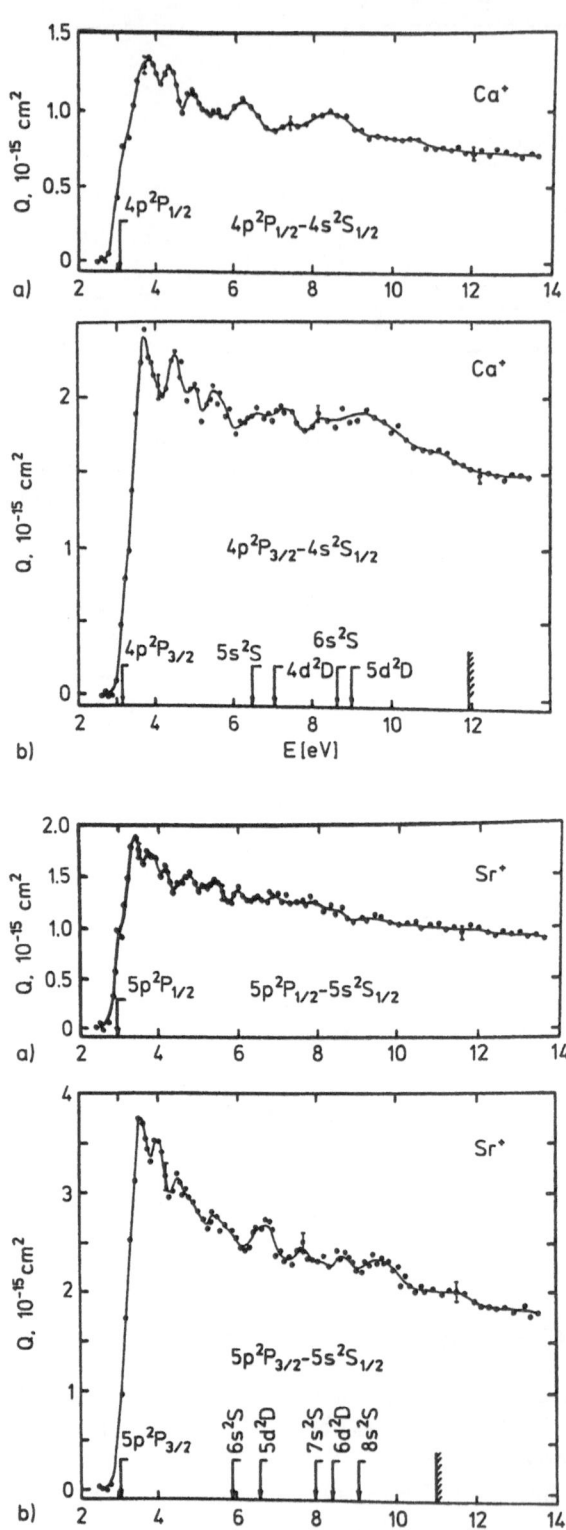

Fig. 7.34a,b. Total cross sections for electron-impact excitation of the components of the 4p $^2P_{1/2,3/2}$ level of Ca$^+$. Experiment: (•) [7.150], (—) digital filtering of the experimental points

Fig. 7.35a,b. Total cross sections for electron-impact excitation of the components of the 5p $^2P_{1/2,3/2}$ level of Sr$^+$. Experiment: (•) [7.150] (—) digital filtering of the experimental points

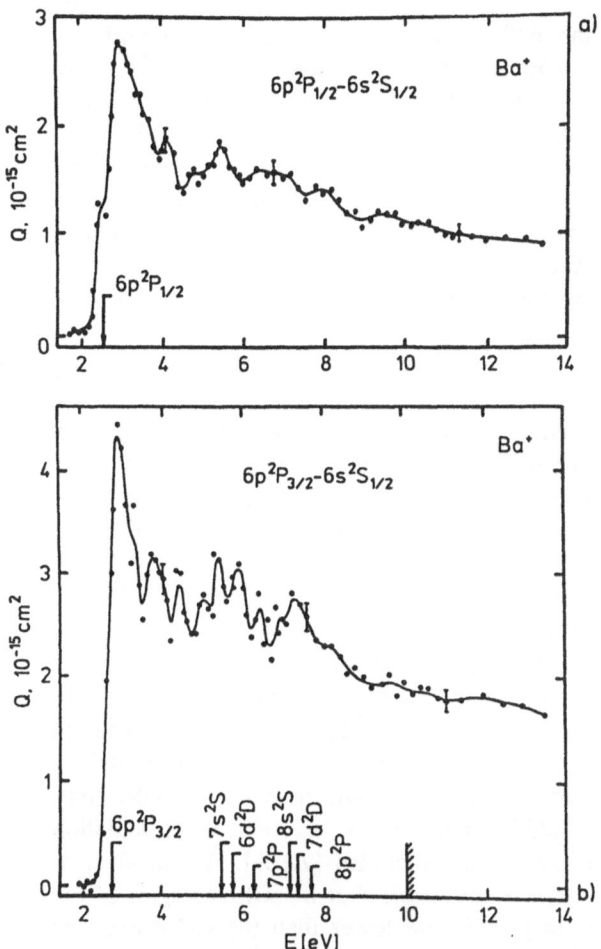

Fig. 7.36. Total cross sections for electron-impact excitation of the components of the 6p $^2P_{1/2,3/2}$ level of Ba$^+$. Experiment: (•) [7.151], (—) digital filtering of the experimental points

mum contribution to the polarizability of the nl state. Usually only a few states of the target, which make the principal contribution to the main polarizability component – the dipole component – are included into the expansion. In our case, the main contribution to the dipole polarizability of the 2p $^2P^0$ state of Be$^+$ is given by the 3d ^2D and 3s ^2S state. Therefore, the $2pnL \pm 1$, $3snL$ and $3dnL \pm 0, 2$ configurations have been included in [7.160] into the basis sets for the $2pnl$ ^{2S+1}L AIS of Be$^+$ which corresponds to the Be$^+$ (2s–2p–3s–3d) CCM. The maximum basis dimension is 25 configurations.

Once the interaction matrix is diagonalized we obtain the energies of the AIS with respect to the ground state of the core. The frozen core approximation, which is calculated within the Hartree-Fock approach, specifies that the correlation energy of the core must be neglected as well as the correlations between the core electrons and valence electrons and the core polarization by valence electrons.

Since the mutual positions of the AIS in question are determined mainly by interaction of valence electrons in Be, the correlation interaction which is not included will result in an energy shift of all calculated AIS. Thus, in order to obtain accurate absolute energy values, it is necessary to know this shift. Usually, the value of this shift is taken to be the difference between the calculated and experimental energies of some reference state of the ion which lies in the energy region studied. Here we shall take the 2s ^{2}S state as the reference state which implies the inclusion of the correlation between core electrons, between core electrons and the 2s electron, and the polarization of the core by 2s electron. Then the energies of 2pnl states are given by

$$E = E_i + \mathcal{E} + E^{\text{exp}}(2s) - E^{\text{calc}}(2s) \tag{7.7}$$

where E_i is the energy of the double ionization of Be (E_i = 27.53 eV), $\mathcal{E}$ is the calculated energy of the AIS relative to the ground state energy of Be^{++}, E^{exp} and E^{calc} are experimental (E^{exp} = -18.211 eV) and calculated (E^{calc} = -18.134 eV), by the Hartree-Fock approximation, energies of the Be$^+$(2s) ion also relative to the ground state.

The characteristics of the Be atom AIS have also been calculated in [7.161–163]. Some of them are listed in Table 7.8. It can be seen from this table that the calculated energies of the AIS of Be are in good agreement with the experimental data. They agree also with the results of the CCM calculations for nine states with core polarization [7.167] except for the 2p^2 ^{1}S and 2p3s ^{1}P^0 states. In the last case, to achieve a better agreement of the results of DM calculations with experiment, one has to extend the basis. As could be expected, DM gives better results than CCM for 2s–2p(3d) states [7.161, 162] and the calculation in which the polarization of the core is not included [7.168]. A typical error in the energy in DM is about 0.03–0.04 eV. Note that the DM energies of the AIS which have been calculated in [7.169] are lower than the CCM energies [7.161] which included more Be$^+$ ion states. As for the widths of the AIS, the discrepancy between the DM [7.169] and 9-state CCM results [7.167] is more significant, typically a factor of 2. Nevertheless, this deviation is not unusually large if one takes into account the sensitivity of the widths to the accuracy of the continuous spectrum functions. For instance, the discrepancy between the 9-state CCM calculations [7.167] and the 3-state CCM calculations [7.160] or between the results of [7.160] and of [7.168] sometimes exceeds a factor of 2 substantially.

The 2pns ^{1}P^0 states of the Be atom have the widest AIS, which fully determine the shape of the photoionization cross section for Be in the 120–84 nm range. These states have very wide symmetric peaks [7.164]. The second series of the ^{1}P^0 AIS has the 2pnd configuration and, as calculations show, has a width which is three times narrower than that for the 2pns series. Due to the anomalous large widths caused by the coupling with the continuum, the 2pns ^{1}P^0 AIS has a significant energy shift Δ relative to its unperturbed position. This shift is related to the widths Γ by $\Delta \approx [\ln 3/2\pi]\Gamma$ [7.170]. In the case of narrow resonances this shift can be neglected but in the case of very wide resonances it gives an

Table 7.8. Energies [eV] and widths [eV] of the AIS of the Be atom

Classification	Theory				Experiment
	[7.160] b	7.161] c	7.162 c	[7.163] d	[7.164–166] e
$2p^2$ ^1S	9.74 (1.75 − 2)	10.20 (1.21 − 3)	9.65 (1.05 − 2)	−	−
2p3p	11.88 (2.26 − 3)	11.92 (2.66 − 3)	−	−	−
2p6p	12.95 (2.04 − 5)	−	−	−	−
2p7p	13.04 (1.36 − 5)	−	−	−	−
2p3p ^3S	11.31 (2.25 − 2)	11.38 1.95 − 3)	11.36 1.03 − 2)	−	−
2p6p	12.88 (3.06 − 3)	12.89 (1.26 − 3)	−	−	−
2p7p	13.00 (1.66 − 3)	−	−	−	−
2p3s $^1P^0$	10.73 (1.074)	10.99 (3.63 − 1)	10.96 (5.01 − 1)	10.77	10.93
2p3d	11.85 (4.71 − 3)	11.92 (2.68 − 4)	−	11.86 (8.08 − 4)	11.85
2p4s	12.09 (2.81 − 1)	12.13 (1.30 − 1)	−	12.07	12.10
2p4d	12.49 (1.94 − 3)	12.51 (2.95 − 5)	−	12.47 (2.63 − 3)	12.50
2p5d	12.78 (9.48 − 4)	−	−	−	12.79
2p6s	12.83 (6.16 − 2)	−	−	−	12.81
2p6d	12.94 (5.49 − 4)	−	−	−	12.95
2p3s $^3P^0$	10.62 (2.30 − 4)	10.65 (1.99 − 2)	10.63 (3.00 − 4)	10.64 (1.97 − 4)	10.61[a]
2p3d	11.80 (1.88 − 2)	11.87 (7.51 − 3)	−	11.81	11.81[a]

[a] Calculated in [7.166];
b – DM; c – CCM; d – Fano's method; e – photoionization

Table 7.8. Continued

Classification	Theory [7.160]	[7.161]	[7.162]	Classification	Theory [7.160]	[7.162]
2-5d ³P⁰	12.76 (4.85 – 3)	–	–	2p6f ³D	12.93 (1.08 – 6)	–
2p6s	12.82 (2.57 – 4)	–	–	2p7p	12.99 (7.13 – 5)	–
2p6d	12.93 (2.80 – 3)	–	–	2p3d κ¹F⁰	11.84 (9.22 – 2)	11.84 (9.05 – 2)
2p7s	12.97 (1.68 – 4)	–	–	2p4d	12.48 (4.23 – 2)	–
2p7d	13.03 (1.91 – 3)	–	–	2p5g	12.77 (9.10 – 6)	–
2p3p ¹D	11.44 (1.79 – 1)	11.51 (1.26 – 1)	11.46 (1.12 – 1)	2p5d	12.78 (2.25 – 2)	–
2p4p	12.29 (6.66 – 2)	12.35 (4.76 – 2)	–	2p6g	12.93 (1.10 – 5)	–
2p6p	12.87 (1.81 – 2)	–	–	2p6d	12.94 (1.37 – 2)	–
2p6f	12.94 (1.14 – 4)	–	–	2p3d ³F⁰	11.66 (1.62 – 1)	11.66 (1.52 – 1)
2p7p	13.00 (1.11 – 2)	–	–	2p4d	12.39 (5.28 – 2)	–
2p3p ³D	11.18 (5.49 – 3)	11.21 (2.10 – 3)	11.18 (2.07 – 3)	2p5d	12.73 (2.69 – 2)	–
2p4p	12.24 (9.70 – 4)	12.24 (7.84 – 4)	–	2p5g	12.77 (2.60 – 6)	–
2p5f	12.77 (2.79 – 6)	–	–	2p6d	12.91 (1.48 – 2)	–
2p6p	12.87 (1.52 – 4)	–	–	2p6g	12.93 (9.03 – 7)	–

appreciable correction to the energy of the AIS. Thus, the calculations of the energies of the $2p^2$ ¹S, 2p3s ¹P states in [7.160] differ from the calculations in [7.161, 162–167] where this shift was taken to be approximately 0.2 eV, but agree remarkably with the results of the Fano's method [7.163], where this shift was not included.

Unlike the ¹P⁰ AIS of Be, where the 2pns state is 10^3 times as wide as the 2pnd state, the 2pns ³P⁰ states are considerably narrower than the 2pnd ³P⁰ states and the 2p3s ³P⁰ state also has extremely small width. An upper limit for the width of 0.0002 eV has been measured for the 2p3s ³P⁰ AIS in the experiment of Pashen and Kruger in early 20's. The calculated width of this state obtained by CCM is about 10^2 times smaller. It has been suggested that to obtain the width of the 2p3s $^3P^0$ state, one must use a very exact wave function for this state [7.171]. A multiconfiguration wave function of this AIS accomodates both angular and radial electron correlations quite well [7.160] and yields a width of 0.00023 eV. This agrees rather well with experimental results and with the latest exact CCM calculations [7.162], 0.00030 eV.

To calculate the 3pnl ^{2S+1}L-AIS of the Mg atom, the 3p$nL \pm 1$, 3snL and 3d$nL \pm 0, 2$ configurations were included into the basis sets as in the case of the Be atom [7.172]. Thus, the dipole polarizability of the 3p $^2P^0$ state of Mg$^+$ was taken into account by including the 3dnl and 4snl configurations into the basis.

The energies and the widths of some 3pnl AIS of Mg calculated in [7.172] by the DM are compared in Table 7.9 with the experimental data and with the results of calculations by the CCM, Fano's method and by the single-configuration Hartree-Fock method. We see that the energies of the AIS calculated by the DM are in good agreement (especially of the high-lying ones) both with the experimental energies and with the energies obtained by CCM. However, the energies of lower AIS in most series lie below the experimental values. It is evident that using the 3p $^2P^0$ level of Mg$^+$ as the matching point, the reference state, results in an overestimation of the correlation and polarization effects in the lower AIS and the energies are too low. For higher AIS ($n \gg 1$) the nl electron is removed far from the core and, hence, the polarization of the core and the correlations of valence electrons with core electrons are dependent mainly only on the 3p electron. Therefore, the choice of the 3p $^2P^0$ level as the reference state allows one to obtain good agreement of the calculated AIS energies with experimental values. In the cases of low-lying AIS, however, the nl electron approaches the core and pushes the 3p electron into the region of large r. Thus, the correlation interaction between the core electrons is reduced to the polarization of the core by the 3p electron.

Recent calculations of the AIS parameters [7.173, 178] have shown that, as in the case of Be, DM calculations [7.172] are in good agreement with other calculations in the resonance energy but exceed them (by about 50 %) in width.

The Be AIS which are attributed to the resonances in the electron excitation of the 2p $^2P^0$ level of Be$^+$, have been calculated using the 3snL, 3p$nL \pm 1$ and 3d$nL \pm 0, 2$ basis configurations [7.179], which corresponds to the (2s–2p–3s–3p–3d) states of Be$^+$ for CCM. The maximum basis dimension in these calculations was 35 configurations. Including the 3pnl and 3dnl configurations ensures including both the dipole and the quadrupole polarizabilities of the 3s ^2S state by the 3p $^2P^0$ and 3d ^2D states as well as the dipole polarizability of the 3p $^2P^0$ state. To obtain the energies of the 3snl, 3pnl and 3dnl AIS (Table 7.10), the 3s ^2S, 3p $^2P^0$ and 3d ^2D states of Be$^+$ were used as the reference states. The energies of the 3snp $^1P^0$ states of Be were calculated in [7.152, 179] and are compared in Table 7.11 with energies measured in photoabsorption experiments [7.164, 165]. We see from Table 7.11 that the energies and the effective quantum numbers n_{eff} of the 3s3p and 3s4p $^1P^0$ states agree remarkably with experimental data. This confirms the fact that the basis configuration set used for the diagonalization takes exact account of the correlation interaction between the valence

Table 7.9. Energies [eV] and widths [eV] of the 3pnl AIS of the Mg atom

Classification	Theory				Experiment		
	[7.172] b	[7.174] c	[7.153] c	[7.173] c	[7.175, 176] d	[7.177] e	[7.175] f
3p² ¹S	8.29 (5.16 – 2)	8.57	8.67 (1.70 – 2)	8.31 (3.47 – 2)	–	8.45	–
3p4p	10.66 (6.08 – 3)	10.70	10.82 (3.40 – 4)		10.93	10.52	–
3p7p	11.71 (1.54 – 3)	–	–	–	11.81	11.70	–
3p4p ³S	11.36 (1.13 – 3)	10.39	–	–	10.79	–	–
3p7p	11.68 (2.93 – 4)	11.68	–	–	11.79	–	–
3p4s ¹P⁰	9.63 (5.85 – 1)	9.70	9.75 (4.30 – 1)		9.90[a]	9.81	9.75
3p3d	19.60 (1.69 – 2)	10.66	10.69 (4.80 – 3)		10.95[a]	10.64	10.65
3p5s	10.91 (1.45 – 1)	10.91	10.93 (1.24 – 1)		10.97[a]	10.97	10.92
3p7s	11.61 (3.16 – 2)	11.61	–		11.63[a]	11.64	11.61
3p4s ³P⁰	9.54 (7.40 – 3)	9.59	9.70 (2.90 – 3)	–	9.70[a]	9.54	9.54
3p5s	10.87 (3.10 – 3)	10.87	10.91 (3.00 – 3)	–	10.91[a]	10.85	10.86
3p6s	11.36 (1.70 – 3)	11.36	–	–	11.38[a]	11.35	11.36
3p7s	11.60 (1.01 – 3)	11.60	–	–	11.61[a]	11.60	11.60
3p4p ¹D	10.54 (3.24 – 1)	10.46	–	–	10.28	10.71	–
3p5p	11.22 (1.26 – 1)	11.19	–	–	11.08	11.31	–
3p6p	11.53 (6.14 – 2)	11.51	–	–	11.39	11.58	–

[a] Calculated in [7.175];
b – DM; c – CCM; d – HF; e – electronic spectroscopy; f – photoionization

Table 7.9. Continued

Classification	Theory			Experiment	
	[7.172]	[7.174]	[7.176]	[7.177]	[7.175]
3p4p ^{3}D	10.24 (2.34 − 2)	10.23	10.19	−	−
3p5p	11.11 (1.05 − 2)	11.10	11.05	−	−
3p6p	11.47 (5.48 − 3)	11.47	11.38	−	−
3p3d ^{1}F^0	10.63 (1.04 − 1)	−	10.75	−	−
3p4d	11.26 (5.02 − 2)	−	11.14	−	−
3p5g	11.53 (9.19 − 6)	−	−	−	−
3p5d	11.55 (3.00 − 2)	−	11.39	−	−
3p6d	11.71 (1.80 − 2)	−	11.65	−	−
3p7d	11.80 (1.40 − 2)	−	−	−	−
3p3d ^{3}F^0	9.95 (2.64 − 1)	−	9.95	9.995	10.03
3p4d	11.04 (4.68 − 2)	−	10.86	11.03	−
3p5d	11.45 (1.96 − 2)	−	11.26	11.44	−
3p5g	11.53 (6.26 − 7)	−	−	−	−
3p6d	11.65 (1.03 − 2)	−	11.44	11.65	−
3p7d	11.76	−	−	11.77	−

electrons giving accurate relative energy values. In addition the choice of the 3s ^{2}S level as the reference state for this series provides the exact absolute energies.

The partial widths of the Be AIS are qualitatively different from those for He [7.87]. This is because of the possibility for the Be atom AIS to decay via the elastic channel. On average, as can be seen from Table 7.10, Γ_1, the width of the decay of AIS to the ground state of the Be$^+$ ion, is nearly 25 % of the total width Γ, although in some cases it has a considerably larger value. For example, for the 3sns ^{1}S AIS, the yield ω_1 is equal to $(\Gamma_1/\Gamma)100 \% \approx 75 \%$. For the 3s5d ^{3}D state, the presence of a neighbouring perturbing AIS, the 3p4p ^{3}D, results,

Table 7.10. Energies E, effective quantum numbers n_{eff}, widths I and yields ω_1 for 3snl, 2pnl and 3dnl AIS of the Be atom [7.179]

State	E [eV]	n_{eff}	Γ [eV]	ω_1 %	State	E [eV]	n_{eff}	Γ [eV]	ω_1 %
3s² ¹S	16.40	1.88	8.18 – 2	79.1	3s8p ³P⁰	20.00	7.25	4.07 – 3	32.2
3p²	18.57	2.24	1.16 – 1	6.2	3s3d ¹D	17.56	2.24	2.20 – 1	31.8
3s4s	18.74	2.99	3.58 – 2	78.0	3p²	18.67	2.29	2.30 – 2	26.8
3s5s	19.45	4.02	1.67 – 2	75.7	3s4d	19.09	3.41	3.89 – 2	48.3
3s6s	19.75	5.03	8.84 – 3	73.4	3s5d	19.56	4.42	1.28 – 2	25.7
3s7s	19.92	6.03	5.27 – 3	71.0	3d²	19.63	2.71	7.96 – 2	4.1
3s4s ³S	18.55	2.82	1.06 – 3	36.3	3s6d	19.79	5.40	3.38 – 3	7.6
3sd5d	19.35	3.87	6.67 – 4	33.9	3s3d ³D	17.78	2.34	3.09 – 2	22.3
3s6d	19.69	4.89	3.27 – 4	36.6	3s4d	19.07	3.38	3.90 – 3	55.3
3s7s	19.87	5.89	2.86 – 4	21.3	3s5d	19.55	4.36	5.31 – 4	92.6
3s3p ¹P⁰	17.68	2.30	1.69 – 1	25.0	3p4p	19.75	2.80	3.32 – 3	1.4
3s4p	18.83	3.09	3.231 – 2	17.1	3s6d	19.85	5.74	4.95 – 3	15.4
3s5p	19.41	4.01	6.25 – 3	10.3	3s7d	19.94	6.56	1.51 – 3	29.8
3s6p	19.68	4.84	1.61 – 2	33.0	3p3d ¹F⁰	18.95	2.32	2.14 – 2	25.9
3p4s	19.77	3.00	3.35 – 2	1.2	3s4f	19.43	4.04	1.55 – 2	19.5
3s7p	19.82	5.58	2.82 – 2	30.5	3s5f	19.70	4.90	7.17 – 3	17.2
3s8p	19.93	6.38	1.43 – 2	30.0	3s6f	19.85	5.76	2.35 – 3	18.0
3s3p ³P⁰	16.96	2.03	9.01 – 2	29.8	3s7f	19.94	6.52	2.31 – 4	54.4
3p3d	18.86	2.37	4.60 – 2	3.8	3p3d ³F⁰	18.30	2.07	1.40 – 2	24.3
3s4p	18.97	3.25	2.63 – 2	51.1	3s4f	19.29	3.75	1.50 – 3	72.1
3s5p	19.50	4.23	1.19 – 2	40.9	3s5f	19.66	4.76	6.53 – 4	58.5
3p4s	19.63	2.86	1.80 – 2	13.2	3s6f	19.85	5.76	3.45 – 4	37.0
3s6p	19.77	5.25	6.44 – 3	37.4	3s7f	19.96	6.71	2.59 – 4	7.1
3s7p	19.91	6.24	3.79 – 3	32.2					

Table 7.11. Comparison of theoretical and experimental energies E [eV] and effective quantum numbers n_{eff} of the 3snp ¹P⁰ AIS of the Be atom

State	Theory		Experiment		
	E [7.152]	n_{eff} [7.152]	E [7.164]	E [7.165]	n_{eff} [7.165]
	a	a	b	b	b
3s3p	17.68	2.30	17.64	17.68	2.30
3s4p	18.83	3.09	18.80	18.83	3.09
3s5p	19.41	4.01	–	–	–
3s6p	19.68	4.48	–	–	–
3s7p	19.82	5.58	19.81	19.80	5.42
3s8p	19.93	6.38	19.93	19.91	6.28
3s9p	20.00	7.33	–	19.99	7.17

a – DM; b – photoionization

Table 7.12. Energies E, effective quantum number n_{eff}, widths Γ and yields ω_1 for $4snl$, $3dnl$ and $4pnl$ AIS of the Mg atom [7.180]

State	E [eV]	n_{eff}	Γ [eV]	ω_1 %	State	E [eV]	n_{eff}	Γ [eV]	ω_1 %
$4s^2\,^1S$	13.04	2.04	1.01 – 1	46.3	$4s3d\,^1D$	13.66	2.27	2.72 – 1	25.6
$3d^2$	14.66	2.71	5.02 – 2	47.2	$3d^2$	14.38	2.50	2.69 – 1	2.8
4.5.	14.97	3.20	4.73 – 2	32.7	$4s4d$	14.96	3.19	1.89 – 2	11.2
$4s6s$	15.53	4.20	1.85 – 2	70.1	$3d5s$	15.30	3.30	9.51 – 2	20.4
$3d4d$	15.64	3.96	1.29 – 2	9.7	$4p^2$	15.49	2.51	5.78 – 2	3.5
$4s7.$	15.80	5.23	1.07 – 2	52.4	$3d4d$	15.55	3.69	8.76 – 2	1.6
$4s5s\,^2S$	14.78	2.99	3.56 – 3	15.2	$4s5d$	15.66	4.49	2.48 – 2	15.5
$3d4d$	15.40	3.50	4.22 – 4	10.7	$4s3d\,^3D$	13.56	2.23	1.03 – 1	11.8
$4s6s$	15.47	4.05	2.12 – 3	14.0	$4s4d$	15.04	3.29	1.90 – 3	71.9
$4s7s$	15.77	5.07	8.84 – 4	22.5	$3d5s$	15.26	3.30	2.72 – 2	11.3
$4s4p\,^1P^0$	14.18	2.54	1.43 – 1	28.2	$3d4d$	15.34	3.41	1.65 – 3	19.4
$3d4p$	14.95	2.95	1.62 – 1	0.5	$4s5d$	15.60	4.42	2.45 – 3	57.9
$4s5p$	15.29	3.68	3.01 – 2	33.3	$3d4p\,^1F^0$	14.66	2.71	2.30 – 1	30.1
$3d5p$	15.56	3.79	7.58 – 2	0.6	$4s4f$	15.28	3.65	1.13 – 2	11.4
$4s6p$	15.64	4.53	6.67 – 3	42.9	$3d5p$	15.53	3.73	5.89 – 2	30.0
$3d4f$	15.74	4.21	4.48 – 2	1.5	$3d4f$	15.63	3.93	5.35 – 3	21.6
$4s7p$	15.86	5.55	4.76 – 3	36.0	$4s5f$	15.71	4.82	2.05 – 2	22.7
$4s4p\,^3P^0$	13.76	2.31	6.07 – 2	11.2	$3s6p$	15.88	4.64	1.09 – 2	30.3
$3d4p$	14.71	2.75	4.59 – 2	3.6	$4s6p$	15.90	5.85	1.31 – 2	30.4
$4s5p$	15.19	3.49	2.01 – 2	22.1	$3d4p\,^3F^0$	14.40	2.54	4.35 – 2	19.8
$3d5p$	15.52	3.71	3.69 – 3	6.0	$4s4f$	15.26	3.61	1.04 – 2	6.9
$4s6p$	15.63	4.50	5.80 – 3	17.9	$3d4p$	15.50	3.67	1.58 – 2	12.7
$3d4f$	15.68	4.05	9.56 – 3	9.6	$3d4f$	15.57	3.81	4.08 – 3	4.6
$4s7p$	15.86	5.59	4.03 – 3	25.9	$4s5f$	15.69	4.70	4.46 – 3	17.6
$3d6p$	15.90	4.71	1.80 – 3	1.9	$3d6p$	15.88	4.66	6.72 – 3	4.7

first, in a sharp reduction of the total width Γ in comparison with other $3snd\ ^3D$ AIS and, second, Γ_1 is almost equal to the total width, $\Gamma_1 \approx 0.93\Gamma$.

The $4snL$, $3dnL \pm 0,2$, $4pnL \pm 1$ basis configurations have been used in [7.180] to caclulate the $4snl$, $3dnl$ and $4pnl$ AIS of the Mg atom. The maximum dimension of the basis was 34 configurations. To obtain the absolute energies (Table 7.12), the $4s\ ^2S$, $3d\ ^2D$ and $4p\ ^2P^0$ states of Mg⁺ were used as reference states. Until now only the optical $4snp$ AIS of Mg have been studied both theoretically and experimentally. The energies of these AIS calculated by DM in [7.152, 180] are compared in Table 7.13 with experimental data and with the theoretical values calculated by the CCM and single-configuration Hartree-Fock approximation.

It can be seen from Table 7.13 that the energy values calculated in [7.152] are in better agreement with the experimental data than those obtained by the CCM or within the Hartree-Fock approximation. This suggests that the correlation interaction between the valence electrons and core polarizability is taken into account more exactly in DM calculations.

The accuracy of CCM depends, as is known, mainly on the accuracy of the target wave functions and on the convergence of the total wave function

Table 7.13. Theoretical and experimental energies E [eV] of the 4snp ^{1}P^0 AIS of the Mg atom

State	Theory				Experiment			
	DM 34×34 [7.152]	DM 21×21 [7.180]	CCM [7.155]	Hartree-Fock [7.175]	a [7.181]	b [7.164]	b [7.163]	b [7.175]
4s4p	14.18	14.37	14.45	14.98	14.20	14.17	14.18	14.18
4s5p	15.29	15.36	15.38	15.40	15.26	15.27	15.26	15.24
4s6p	15.64	15.70	15.71	15.72	15.66	15.65	15.62	15.61
4s7p	15.86	15.88	15.89	15.90	15.83	–	15.83	15.83
4s8p	15.99	16.00	16.00	16.01	15.95	–	–	15.98

34, 21 Number of configurations;

a – electronic spectroscopy; b – photoionization

expansion. It has been noted in [7.155] that although the target orbitals used might ensure good accuracy, the expansion does not converge and, hence, the calculated energies of the 4snp ^{1}P^0 resonances considerably differ from the experimental values. The reason is that in the case of the (3s–4p–4s–3d) coupled channels considered in [7.155] only the quadrupole polarization of the 4s ^{2}S state by the 3d ^{2}D state was taken into account while the more important dipole polarization was not included. This polarization was accounted for in the calculations of [7.180] by introducing the 4pns and 4pnd configurations into the basis. Excellent agreement with the experiment was obtained. The calculations were repeated with a limited 21×21 basis without these configurations to evaluate the influence of 4pns, nd configurations. The resulting energies, reported in Table 7.13, lie higher than those obtained when these configurations are included and are close to the data of [7.155]. Note that the effect of polarization of the 4s ^{2}S state by the 4p ^{2}P^0 state is most pronounced for the 4s4p ^{1}P^0 state where the energy is decreased by approximately 0.2 eV.

These calculations reveal a considerable configuration mixing for the ^{1}P^0 state. For example, the 4s4p basis configuration makes up only 55% of the contribution to the wave function of the lower 4s4p ^{1}P^0 AIS. Therefore, the single-configuration Hartree-Fock approximation leads to considerably overestimated energy values for these AIS. For instance, the Hartree-Fock energy of the 4s4p ^{1}P^0 state [7.175] is higher than the experimental one by 0.8 eV. For higher members of a series for which the importance of CI decreases, the agreement of the Hartree-Fock energies with experimental data improves.

Similarly to the Be atom, the partial width Γ_1 of the AIS of the Mg atom is a considerable fraction of the total width Γ. This is seen most clearly for the 4snl AIS. The AIS with 3dnl configurations decay mainly into the excited 3p ^{2}P^0 state of Mg$^+$ rather than into the ground 3s ^{2}S state. So, the probability of the 3dnp, nf ^{1}P^0 state decaying into the 3s ^{2}S state is only a few percent of the total decay probability, while for the 4snp ^{1}P^0 AIS this reaches 40%. However, in general, the AIS decay for Be and Mg through autoionization occurs primarily into the nearest low-lying excited states of the corresponding ion.

Unlike the AIS of a He atom for which the decay probability of the AIS which converge to the $n = 3$ states of He^+ into the 1s state of He^+ is extremely small, the probability of decay of the $3(s,p,d)nl$ states of Be and $4(s,p)nl$, $3dnl$ states of Mg into the ground state of the ion is a considerable part of the total decay probability. In some cases, it reaches more than 90%.

7.6.2 Resonances in Elastic e + Be⁺ and e + Mg⁺ Scattering

To study the influence of electron capture into the $2pnl$ state of Be and $3pnl$ state of Mg in elastic electron scattering by Be^+ and Mg^+ ions, we have calculated, by DM, the elastic differential cross sections in the energy range from 0.1 eV to the excitation thresholds of the 2p $^2P^0$ Be^+ (3.96 eV) and 3p ^2P Mg^+(4.42 eV) states [7.153]. The nonresonant part of the scattering amplitude was found by a single-channel version of the *IMPACT* routine [7.121]. Table 7.14 presents the S, P, D and F singlet and triplet phase shifts of the elastic electron-beryllium ion scattering at electron energies from $k^2 = 0.00735$ Ry to $k^2 = 0.40$ Ry obtained by a numerical solution of (6.82). Only the first five partial waves were included into the summation over L since for $L \geq 5$ the scattering phases have values smaller than 10^{-4} rad.

Table 7.14. Phase shifts δ_L [radian] for elastic scattering of electrons by the Be^+ ion [7.152]

k^2 [Ry]	^1S	^3S	$^1P^0$	$^3P_0$	^1D	^3D	$^1F^0$	$^3F^0$
0.0	2.100	2.437	1.174	1.120	−0.320	0.343	0.111	0.112
0.00735	1.739	2.292	0.203	1.005	−0.045	0.156	0.001	0.009
0.02205	1.736	2.286	0.223	0.997	−0.045	0.162	0.002	0.012
0.05880	1.729	2.270	0.267	0.977	−0.046	0.175	0.002	0.016
0.07350	1.727	2.263	2.284	0.970	−0.047	0.179	0.003	0.017
0.14701	1.716	2.231	0.353	0.936	−0.044	0.201	0.004	0.027
0.2	1.714	2.212	0.396	0.918	−0.043	0.212	0.005	0.032
0.3	1.711	2.175	0.457	0.887	−0.037	0.230	0.009	0.047
0.4	1.710	2.143	0.501	0.861	−0.028	0.244	0.013	0.060

It can be seen from Table 7.14, that the long-range Coulomb potential in electron-ion scattering leads to different behaviour of the phase shift as $k^2 \rightarrow 0$ from that observed in electron-neutral scattering. Indeed, in the case of elastic electron scattering by a neutral system the scattering phase shifts obey the relation

$$\tan \delta_l \xrightarrow[k^2 \rightarrow 0]{} \text{const } k^{2l+1} , \tag{7.8}$$

and, hence, it follows that for all l the scattering phase shifts vanish at the threshold. In the case of electron-ion scattering the phase shifts have finite values at the threshold which are related to the quantum defects $\mu_l(n)$ of the energy levels with a given angular momentum. These levels lie below the continuous spectrum,

$$\lim_{n \to \infty} \pi \mu_{l(n)} \to \lim_{k^2 \to 0} \delta_l(k^2) \, . \tag{7.9}$$

Using (7.9) one may find the elastic scattering phase shifts at low k^2 starting only from the energy values of the bound atomic states. The first row of Table 7.14 shows the threshold phase shifts calculated by (7.9). The energies of the bound Be states used as reference states were taken from [7.162]. The discrepancies between the threshold values of the scattering phase shifts obtained from the experimental energies and those extrapolated from our calculated values result from the neglect of the polarization of Be^+ and Mg^+ ground states induced by higher excited states in the nonresonance scattering calculatons. Within DM the core polarizability may be taken into account like in CCM by introducing, into the initial system, additional parametric polarization potentials [7.162]:

$$V(\varrho, r) = \frac{\alpha}{r^4} W_6(r/\varrho) \, , \tag{7.10}$$

where $W_n(x) = 1 - \exp(-x^n)$, α is the experimental value of the core polarizability. The fitting parameter ϱ is chosen so that the energies of several lower ionic states which are obtained by solving the Hartree-Fock equations with (7.10) coincide with their experimental values.

As many as 44 AIS of the Be atom were included in the calculation of the elastic electron scattering by the Be^+ ion by (6.88).

Figure 7.37 shows the energy dependence of the differential cross section for the elastic scattering $e + Be^+$ for various scattering angles; we also found the positions of resonances with $\Gamma \geq 10^{-4}$. The results demonstrate that the resonances are manifested most clearly in the large-angle scattering. In contrast to the cross section for photoionization of ground state Be where all the structure is entirely due to broad $2pns\ ^1P^0$ resonances and narrow $2pnd\ ^1P^0$ resonances, the structure in the cross section for the elastic scattering $e + Be^+$ is dominated by triplet resonances of $2pnd\ ^3P$, $2pnp\ ^3D$ and $2pnd\ ^3F^0$ and the singlet $2pnd$ $^1F^0$ resonances. The singlet S, P, and D resonances, on the other hand, make a significantly smaller contribution. At the scattering angle $\theta = 180°$, the $^{1,3}F^0$ resonances give rise to a series of clearly defined peaks. Against the background of these peaks, the $2pnd\ ^3P^0$ and $2pnp\ ^3D$ resonances are seen as narrow minima. The minima caused by the $2pnp\ ^3D$ resonances as well as the contribution of the $2pnp\ ^3S$ and ^1D resonances are seen most clearly at $\theta = 90°$, while the P and F resonances do not contribute to the cross section.

The resonance structure in the differential cross section for the elastic scattering of electrons by the Be^+ ion is significantly richer. It differs qualitatively from the structure in the corresponding cross section for the He^+ ion. This is because the widths of the $2pnl$ AIS of Be are, on the whole, substantially greater than those for the $(2, n\alpha)$ AIS of He, especially if $L \geq 1$. While the widths of the $2pnd\ ^3P^0$ AIS of Be for $L = 1$ are only twice the widths of the $2pns\ ^3P^0$ AIS of He for $L = 2$, the widths of the $2pnp\ ^3D$ AIS of Be are more than an order of magnitude larger than those of the $(2, n\alpha)\ ^3D$ AIS of He. For AIS with $L = 3$,

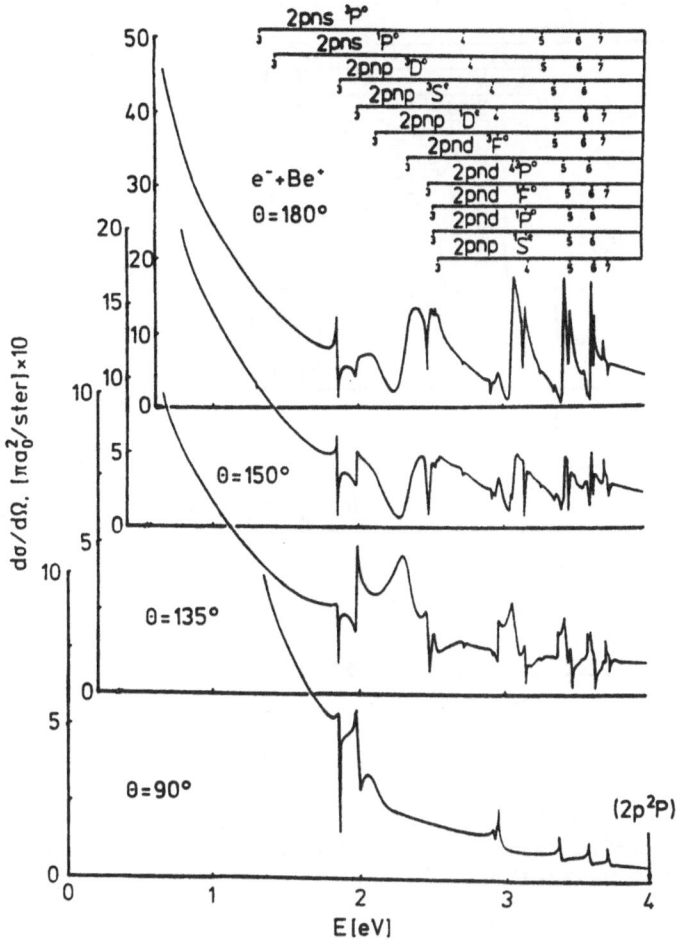

Fig. 7.37. Energy dependence of the differential cross section for elastic scattering of electrons by the Be$^+$ ion for several scattering angles. From [7.152]

which determine the behaviour of the cross section for the elastic scattering of electrons by Be$^+$, this difference increases to two orders of magnitude.

Figure 7.38 shows the energy dependence of the differential cross section for the elastic scattering of electrons by the Mg$^+$ ion; the positions and configurations of the corresponding AIS of Mg are also shown. We see that again the structure in the cross section is primarily due to triplet resonances; seen particularly clearly is a low-lying 3p3d $^3F^0$ resonance with Γ = 0.264 eV. Since the phase shift $\delta_{l=3}^{S=1}$ of the nonresonance scattering at the energy of the 3p3d $^3F^0$ AIS is small ($\sim$ 0.04 rad), these resonances reveal broad, symmetric Breit-Wigner peaks. The lower 3p3d $^1F^0$ AIS also leads to the appearance (at θ = 180°) of a broad (Γ = 0.104 eV) peak with a narrow dip near its top which is caused by the 3p3d $^1P^0$ AIS. Higher 3pnd $^{1,3}F^0$ resonances are seen clearly at θ = 180° as narrow double symmetric peaks. At the same scattering angle, the ^3D resonances give rise to a

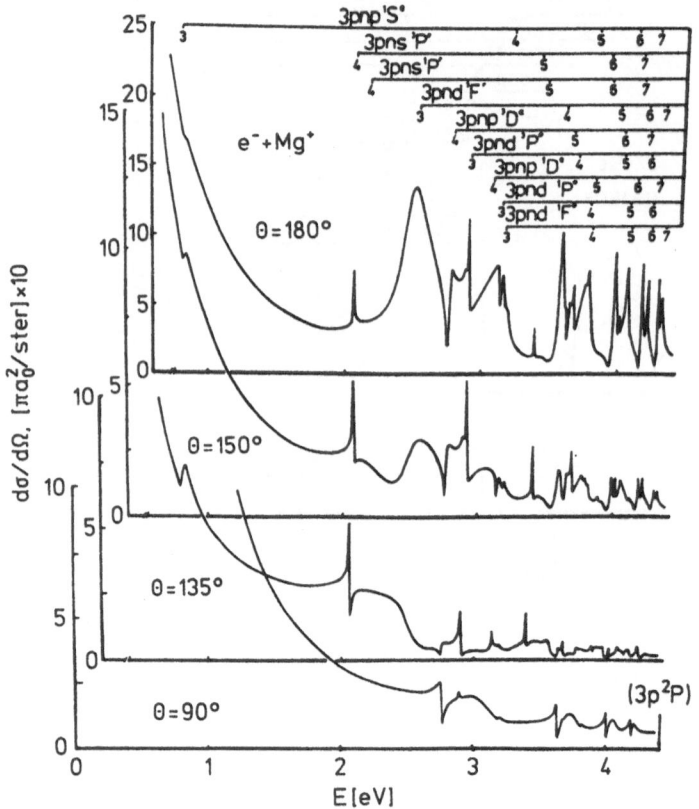

Fig. 7.38. Energy dependence of differential cross section for elastic scattering of electrons by the Mg$^+$ ion for various scattering angles [7.152]

series of narrow dips while the $^3P^0$ resonances, in contrast to e $-$ Be$^+$ scattering, yield the narrow symmetric peaks which are most noticeable at $\theta = 180°$ where the contribution of the F resonances decreases.

The contribution of the very broad $3pns$ $^1P^0$ resonances, which completely determine the shape of the cross section for photoionization of ground-state Mg atom at near-threshold energies, can be detected only at $\theta = 135°$, where the peaks due to the $^{1,3}F^0$ resonances disappear almost entirely. At these scattering angles, the lower $3p4s$ $^1P^0$ resonance ($\Gamma = 0.585$ eV) is seen as a very rounded maximum near 2 eV, the shape of which is very distorted, by a rapid increase of the cross section with energy decrease, $\sim 1/k^4$, and by an additional structure near the top due to the neighbouring $3p4s$ $^3P^0$ resonance.

At $\theta = 90°$, the elastic cross section is dominated by the narrow $3pnd$ ^3D ($\Gamma \leq 0.02$ eV) and broad $3pnp$ ^1D ($\Gamma_{3p^2} \sim 0.3$ eV; $\Gamma_{3p4p} \sim 0.1$ eV; $\Gamma_{3p5p} \sim 0.06$ eV) resonances. Note that the ^3S resonances, which in the case of Be$^+$ are seen at all scattering angles, in the case of Mg$^+$ are observed as weak features only at $\theta = 90°$, since they are an order of magnitude narrower than the resonances in Be$^+$.

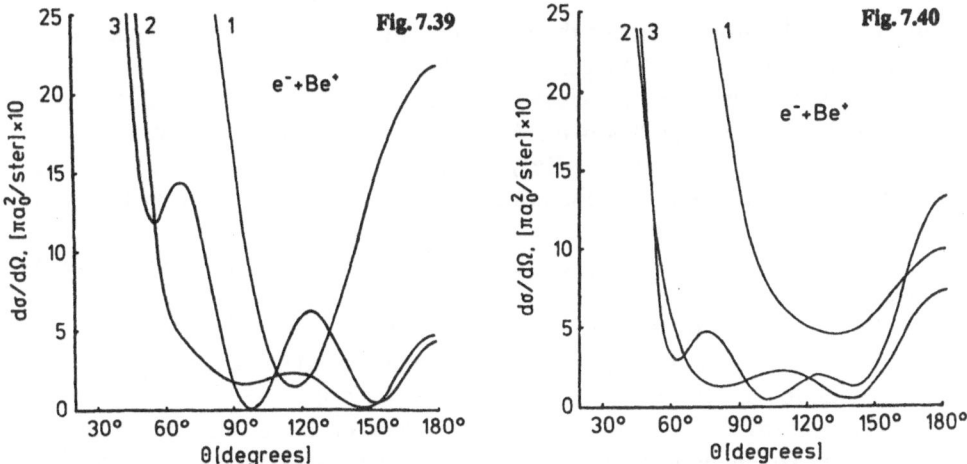

Fig. 7.39. Angular dependence of the differential cross section for elastic scattering of electrons by the Be^+ ion for various energies [7.152]. Curve (*1*) 0.08 Ry; (*2*) 0.169 Ry; (*3*) 0.182 Ry

Fig. 7.40. Angular dependence of the differential cross section for elastic scattering of electrons by the Mg^+ ion for various energies [7.152]. Curve (*1*) 0.08 Ry; (*2*) 0.183 Ry; (*3*) 0.229 Ry

Figures 7.39, 40 show the angular dependences of the elastic differential cross sections for Mg^+ and Be^+ calculated via DM. In the nonresonance region (curve 1), the shapes of the differential cross sections are typical for elastic electron scattering, i.e., the cross section decreases monotoically with increasing scattering angle, reaches a minimum at $\theta \sim 120-135°$ and then increases monotonically up to 180°. In the resonance region, the cross section has a significant asymmetry. Near the lower 3p3d $^3F^0$ resonance (curve 2) the cross section has two local maxima at $\theta \sim 70°$ and $\theta \sim 120°$, respectively, and a deep minimum at $\theta \sim 100°$. These are due to the contribution of a partial amplitude with $L = 3$. In the region of the lower $^3P^0$ resonances (curve 3) the differential cross sections have less structure, with small minima at $\theta \sim 80°$, $\theta \sim 140°$ and with a broad maximum at $\theta \sim 110°$.

The lack of any other (theoretical or experimental) results on these cross sections rules out a comparative analysis. It would be extremely helpful to do an experiment on these processes, since the resonance structure predicted theoretically in the elastic cross sections is very pronounced (particularly in the case of the Mg^+ ion) and can be detected in existing experimental apparatus using intersecting electron and ion beams.

7.6.3 Resonances in Inelastic e + Be⁺ and e + Mg⁺ Scattering

Let us consider in more detail the solution of the DM equations which describe the scattering of electrons at energies high enough to excite low-lying atomic ionic states. This requires the inclusion of the wave functions of all states which may be excited in the first term of the expansion (6.35), in addition to the wave function of the ground state of the ion. Thus, all the channels open at the given energy are included in the first term in (6.35).

To calculate the excitation cross for the $2p\ ^2P^0$ and $3p\ ^2P^0$ levels of Be^+ and Mg^+ in the energy range below the $3s\ ^2S$ and $4s\ ^2S$ threshold, respectively, one must include the $2s\ ^2S$, $2p\ ^2P^0$ (Be^+) and $3s\ ^2S$, $3p\ ^2P^0$ (Mg^+) states. Then the final system of equations for finding $F^0(r)$ will comprise three (two, if $L = 0$) integro-differential equations.

For the Be^+ ion, depending on the $LS\pi$ state of the system, the correlation functions χ_μ with the configurations ^1S–$2s^2$ and $2p^2$, $^{1,3}P^0 - 2s2p$, $^1D - 2p^2$ must be included in (6.35). For other $LS\pi$ states, these correlation functions are omitted. Similar sets of correlation functions must be added to the expansion in calculating the $3p\ ^2P^0$ excitation cross sections for Mg^+.

The method of transformation of a system of linear algebraic integro-differential equations to a system of algebraic equations suggested by *Seaton* [7.182] has been applied in [7.146, 156]. The final system of algebraic equations has been solved using the IMPACT routine [7.121]. The wave functions $F^0(r)$ obtained were used to find the widths Γ according to (6.70), while the K^0-matrix was used to calculate the T^0-matrix of the nonresonance scattering. Note that the experimental energies (0.2909 Ry for Be^+ and 0.3249 Ry for Mg^+) were used as the excitation energies for the $2p\ ^2P^0$ and $3p\ ^2P^0$ levels of Be^+ and Mg^+, respectively.

DM calculations on the partial, total and differential cross sections for the excitation of the resonance levels in Be^+ and Mg^+ ion were done in the energy region from the corresponding thresholds ($k_1^2 = 0.2909$ Ry for Be^+ and 0.3249 Ry for Mg^+) up to the nearest $ns\ ^2S$ levels ($k_1^2 = 0.8040$ Ry for Be^+ and 0.6362 Ry for Mg^+) [7.146, 156]. For the total cross sections 10 lower partial waves have been taken into account in the sum (6.75), the exchange has been taken into account up to $L = 6$.

The calculated partial excitation cross sections for the $2p\ ^2P$ level of Be^+ ($l = 1 - 3, S = 0, 1$) are shown in Fig. 7.41.

It is seen that the resonances are manifested quite clearly in all partial waves and the maximum resonance cross section value (except for the $^1P^0$ and $^1F^0$ states) is 3 to 6 times larger than the nonresonance cross section at the same energy.

The $3snp\ ^1P^0$ resonances which dominate the cross sections for photoionization from the ground state of the Be atom give rise to a slight resonance structure in the corresponding inelastic partial cross section. Taking into account the fact that the absolute value of that partial cross section in the resonance region is small $\sim 0.8\pi a_0^2$, one should not expect a pronounced resonance structure in the total excitation cross section due to the $^1P^0$ resonances.

In the case of the ^3D state, direct excitation is negligible compared to resonance excitaion, thus, interference of the direct $2p\ ^2P^0$ excitation amplitude with that of cascade excitation via the AIS does not occur. As is seen from Fig. 7.41, the shapes of the resonances in the partial cross section resemble narrow totally symmetric Breit-Wigner peaks.

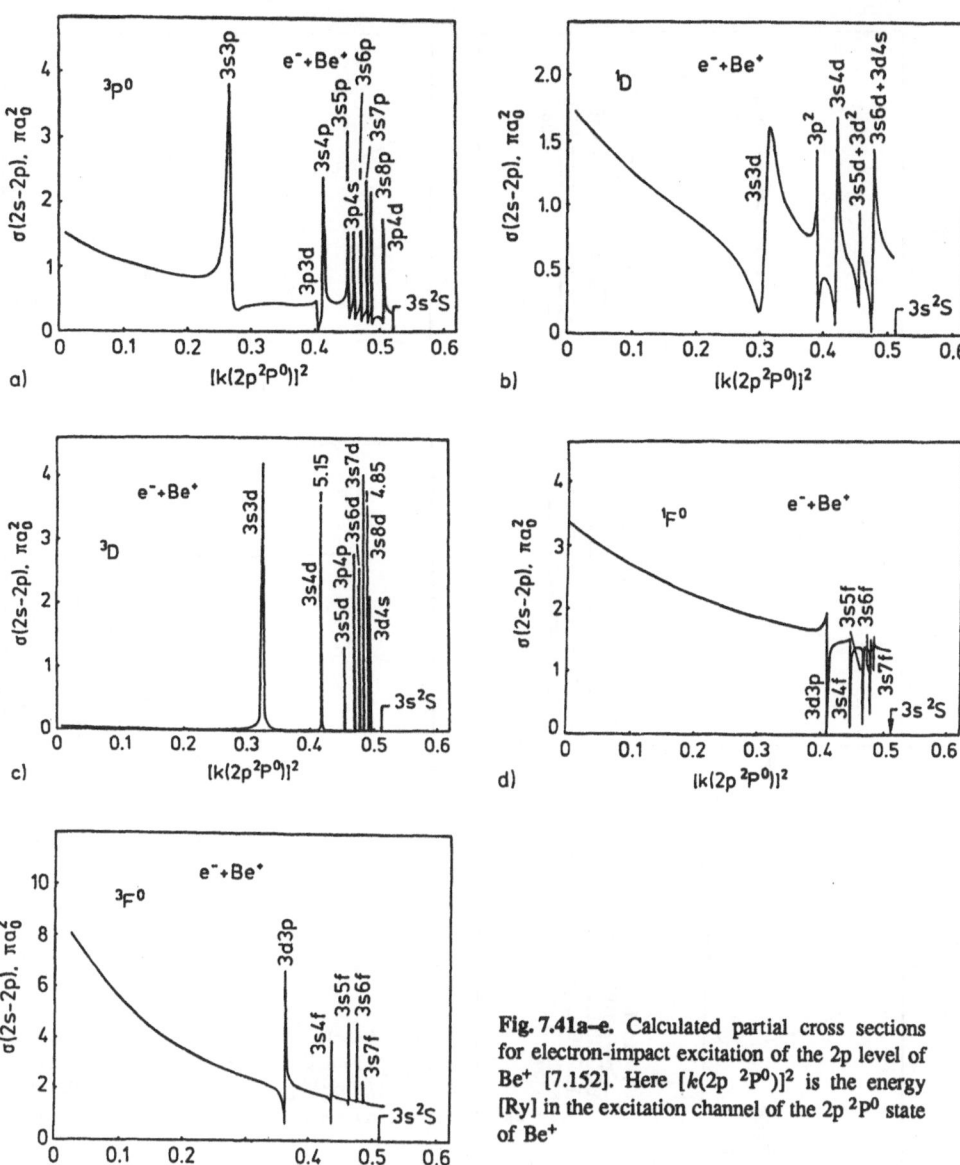

Fig. 7.41a–e. Calculated partial cross sections for electron-impact excitation of the 2p level of Be^+ [7.152]. Here $[k(2p\ ^2P^0)]^2$ is the energy [Ry] in the excitation channel of the 2p $^2P^0$ state of Be^+

The partial excitation cross sections for the 3p $^2P^0$ level of Mg^+ ($L = 1-3$, $S = 0, 1$) are presented in Fig. 7.42. As can be seen, the resonance structure in the partial cross sections for Mg^+ is more complicated than for Be^+. This is due to the near-degeneracy of the 4s ^2S and 3d ^2D levels of Mg^+. Thus, besides the $4snl$ AIS, a large number of $3dn$ AIS converging to the 3d ^2D threshold (50 in the present calculation) are in the region between the 3p $^2P^0$ and 4s ^2S levels. In the case of Be^+, we have found only 5 AIS with $3dnl$ configurations between the 2p $^2P^0$ and 3s ^2S levels.

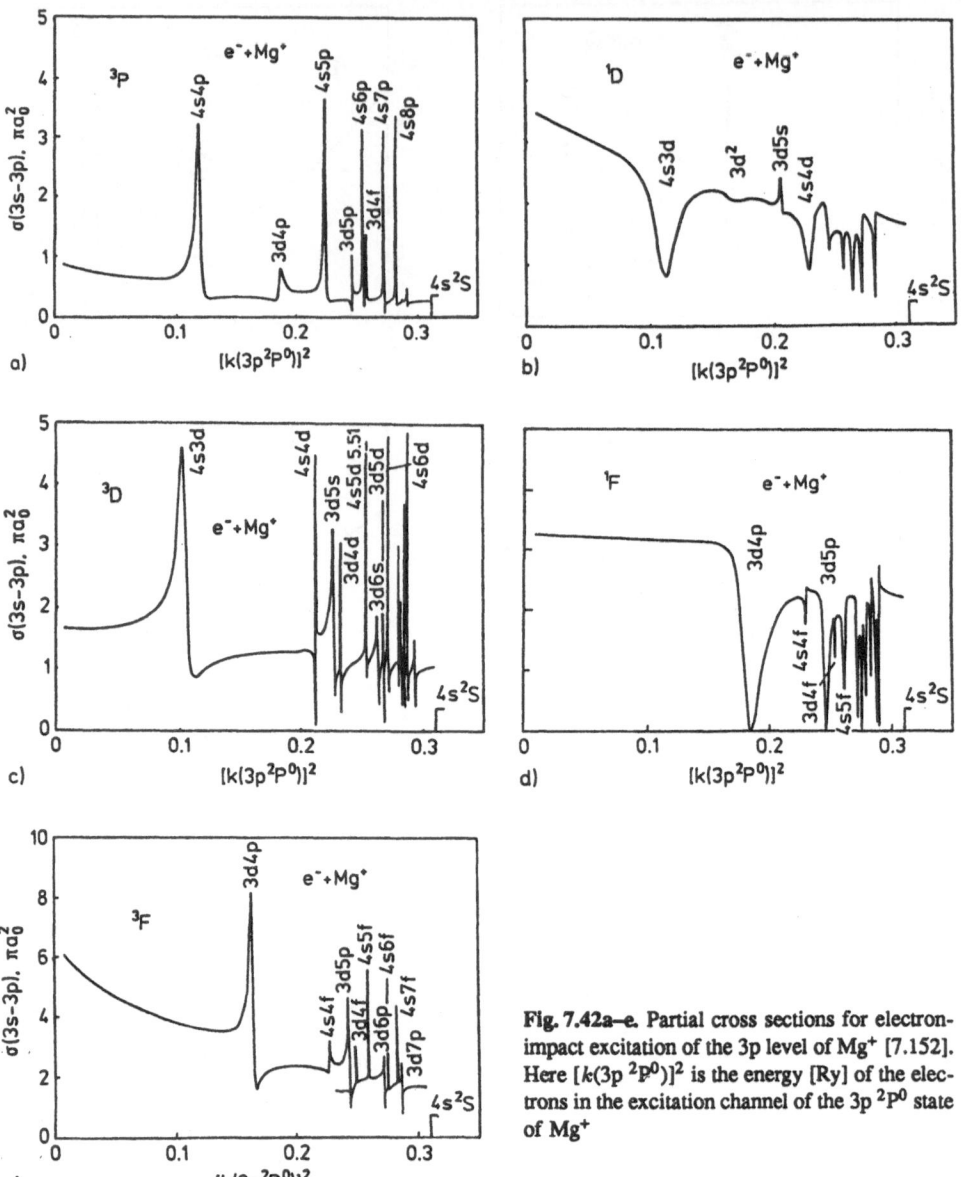

Fig. 7.42a–e. Partial cross sections for electron-impact excitation of the 3p level of Mg+ [7.152]. Here $[k(3p\,^2P^0)]^2$ is the energy [Ry] of the electrons in the excitation channel of the 3p $^2P^0$ state of Mg+

For $L = 0$, the shapes of the $4sns$ and $3dnd$ resonances in Mg are like those of symmetric isolated Breit-Wigner peaks. As for ^3D in Be+, this is due to the small values of the partial amplitudes for the direct excitation process. Once more, similarly to the Be atom, the $^1P^0$ AIS of Mg which are clearly revealed in the photoabsorption cross sections in the vacuum ultra-violet region show a weak resonance structure in the corresponding partial cross section. Due to the small value of the partial $^1P^0$ cross section in the resonance region ($\sim 0.9\pi a_0^2$) the structure in the total cross section is weak too.

Singlet AIS, resulting mainly in the destructive interference of the potential scattering with the resonance scattering are intrinsic features of the partial excitation cross sections for Be$^+$ and Mg$^+$ (and for He$^+$). Triplet AIS induce constructive interference. This is probably a general feature of resonance scattering of electrons by ions with a single valence electron.

The resonance structure in the partial excitation cross sections for the 3p ^{2}P^0 state of Mg$^+$ calculated in [7.146] agrees remarkably with the structure in the partial collision strengths Ω_{LS}(3s $-$ 3p) obtained in [7.155] within the CCM. This means that DM which is, generally speaking, a simplified version of CCM, allows one to take into account the relation between the direct and resonance excitation to good accuracy. DM recreates the multichannel resonance scattering mechanism more clearly and uses the known energy values, partial and total widths of resonances to analyze the resonance structure in the cross sections. For instance, CCM calculations show that the 4s3d AIS causes a very deep minimum in the ^{1}D partial cross section and the minimum related to the next, 3d^2, AIS has a much smaller value. To elucidate the reasons for such differences in the behaviour of the resonances in CCM one needs to perform cumbersome calculations requiring the approximation of the K-matrix by a parametric formula of Breite-Wigner type. Within DM, it is clear that although the 4s3d ^{1}D and 3d^2 ^{1}D AIS have nearly equal widths (0.273 and 0.269 eV, respectively), the 3d^2 ^{1}D AIS will lead to weaker resonance structure in the cross section, since its partial width Γ_1 is nearly 10 times narrower than that for the 4s3d ^{1}D AIS.

Figure 7.32, discussed above, shows the results of calculations of the total cross section for excitation of the resonance level of Be$^+$ along with some experimental results [7.147]. As expected, far from the resonances ($k^2 \leq 0.2$ Ry) the cross section calculated by the DM agrees with that found in [7.157] by the CCM for 2s–2p states of Be$^+$. In the resonance region the cross section exhibits significant structure mostly due to the triplet resonances with $L \geq 1$. At 0.2 Ry $\leq k_2^2 \leq 0.4$ Ry, the lower 3s3p ^{3}P^0, 3s3d ^{3}D and 3p3d ^{3}F^0 AIS of the Be atom give rise to clearly resolved peaks (the 3p3d ^{3}F^0 resonance has the largest relative amplitude). As the energy increases ($k_2^2 < 0.4$ Ry), the resonances become more closely spaced and are seen as a continuous network of very narrow peaks rising by $(3–5)\pi a_0^2$ above the average height of the cross section.

To evaluate the effect of resonances on the excitation cross section under the conditions of an actual experiment, the theoretical cross section was averaged over the energy distribution of the electrons in the beam with a width of 0.3 eV.

As is seen from Fig. 7.32, the mean height of resonances in the average cross section is about 5–10 %, which explains why no resonance structure was observed in the experiments reported in [7.147]. Indeed, since the authors of [7.147] estimate their measurement accuracy to be about 7 %, it is clear that these features would have been within the experimental error.

Figure 7.33 shows the results of DM calculations on the excitation of a resonance level of Mg$^+$, along with the data of a recent experiment [7.146]. The DM cross section averaged over the experimental energy distribution agrees very well both in magnitude and in shape with the experimental cross section after the measurements have been subjected to digital filtering. The calculated cross section is somewhat lower than the calculated cross section, by ~ 6 %, apparently because of imperfect normalization of the experimental cross section to the theoretical cross section found in the Coulomb-Born approximation at 100 eV. In [7.154] detailed CCM calculation of the total cross sections for the Ca^+ 4s–4p and 4s–4d transitions have been carried out. A rather elaborate model which takes into account six states (4s–3d–4p–5s–4d–5p) was used. The comparison of the theoretical predictions [7.154] with the experimental data [7.151] is shown in Fig. 7.34. The many resonance lines predicted seem to be confirmed by experiment. The theoretical calculations of excitation of the 4p level of Ca^+ were also carried out in [7.183] using the DM. The results of calculations are in quite good agreement with the experimental data. Rich resonance structure is predicted by theory below 5s threshold.

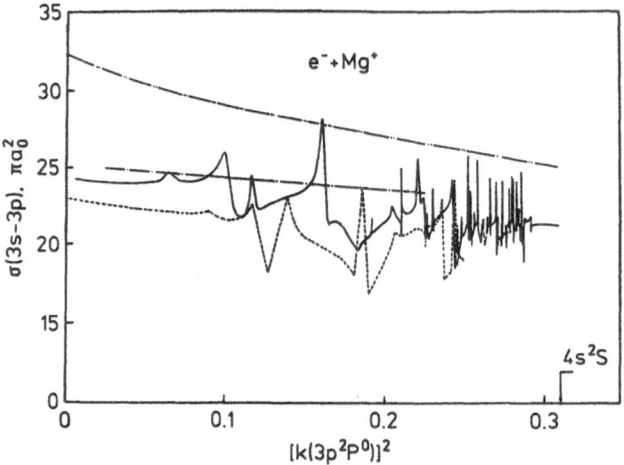

Fig. 7.43. Comparison of calculated total cross section for electron-impact excitation of the 3p level of Mg^+. (—) DM [7.146, 152], (– · –) CCM [7.153], (---) CCM [7.155], (– ·· –) Coulomb-Born approximation [7.158]

Figure 7.43 compares the excitation cross section of Mg^+ calculated by DM with results of other theoretical calculations. We see that the resonance structure found in the excitation cross section by CCM in [7.155] correlates well with the shape of the resonances calculated by DM. For example, both curves reveal peaks due to the 4s3d ^3D, 4s4p $^3P^0$ and 3d4p $^3F^0$ resonances (the most prominent peak is seen at 0.1641 Ry and is due to the 3d4p $^3F^0$ AIS) and the dips due to the 4s3d ^1D and 3d4p $^1F^0$ AIS of the Mg atom. However, despite the fact that the relative positions of the resonances agree remarkably with each other, on the CCM curve these features are shifted about 0.25 eV up. This shift is present because the authors of [7.155] ignored some of the correlations which give rise to the dipole polarization of the 4s ^2S state of Mg^+. In the DM calculations the dipole polarization was taken into account by incorporating the 4pn (s,d) configurations in the basis. It can be seen from Fig. 7.33 that this approach leads to very good agreement between the resonance structure observed experimentally in the excitation cross section of the 3p $^2P^0$ level of Mg$^+$ and that found in DM calculations [7.146].

7.6.4 Electron–Alkali Ion Scattering

We turn now to a brief a nalysis of the ions of alkali elements. Very interesting results on the electron-impact excitation of ions of alkali elements have recently been obtained in [7.184]. A preliminary study has been made of the emission accompanying inelastic collisions of K$^+$, Rb$^+$, and Cs$^+$ with electrons over the energy range of 8 – 400 eV. The structures found here indicate that the interaction is of a resonance nature. *Zapesochny* et al. [7.184] suggest that the decay of certain autoionizing states to the resonance level of the ion occurs without a change in the quantum numbers n_1, l_1 of the excited electron (a Coster-Kronig process)

$$A^+(np^6) + e \rightarrow A^{**} \left[np^5 \left(^2P_{1/2} \right) n_1 l_1 n_2 l_2 \right]$$
$$\rightarrow A^{+*} \left[np^5 \left(^2P_{3/2} \right) n_1 l_1 \right] + e \ .$$

The density of resonance states in the excitation cross section of the ion increases with the decreasing of the excitation energy. A similar situation has been observed in electron excitation of the Tl$^+$ ion [7.185]. Unfortunately, we do not yet have any theoretical calculations which would confirm or refute these observations.

In summary, the status of our understanding of resonances in electron-ion collisions is less advanced than that of electron-atom scattering. The experimental information is scanty. The complete lack of experimental data on differential cross sections prevents the classification of resonances in several cases. The large energy spread of the electron beam used has made it impossible to compare theory and experiment in detail.

7.7 Group II Atoms

We turn now to the scattering of low-energy electrons by alkaline-earth atoms. Following [7.186], where the elastic cross sections and the excitation cross sections for these elements have been calculated, we shall outline the derivation of the close-coupling equations to be used.

Consider two valence electrons of an atom and an incident electron which move in the effective field of the nucleus and the core electrons. In this case the Hamiltonian which describes the system has the form

$$H = \sum_{i=1}^{3} \left[-\frac{1}{2} \Delta_i + V_{core}(r_i) \right] + \sum_{i>j} \frac{1}{r_{ij}} \ , \tag{7.11}$$

where $V_{core}(r_i)$ is the effective potential of the nucleus and the core electrons, $1/r_{ij}$ is the Coulomb interaction between electrons.

The total wave function may be written as

$$\Psi = \sum_i \Big[\Psi_i(1,2)F_i(3)\chi_i(1,2,3) $$

$$+ \Psi_i(2,3)F_i(1)\chi_i(2,3,1) + \Psi_i(3,1)F_i(2)\chi_i(3,1,2) \Big] , \qquad (7.12)$$

where Ψ_i is the coordinate part of the wave function of the valence shell; F_i is that of the incident electron; χ_i is the three-electron spin function. The summation is over all target states to be taken into account. Substituting (7.12) into (6.17) with the Hamiltonian of (7.11), we obtain the system of close-coupling equations

$$\left(\frac{1}{2}\Delta_3 + E - E_i\right) F_i(3) = \sum_i \left[V_{ij}(3)F_i(3) + \int W_{ij}(1,3)F_j(1)dr_1 \right] , \quad (7.13)$$

where $V_{ij}(3)$ and W_{ij} are the direct and exchange potentials, and

$$V_{ij}(3) = \delta_{ij} V_{core}(r_3)$$
$$+ \delta_{s_i s_j} \int \Psi_i^*(1,2) \left(\frac{1}{r_{23}} + \frac{1}{r_{13}}\right) \Psi_j(1,2)dr_1 dr_2 ;$$

$$(7.14)$$

$$W_{ij}(1,3) = b_{ij} \int \Psi_i^*(1,2)(H - E)\Psi_j(2,3)dr_2 ;$$
$$b_{ij} = \chi_i(1,2,3)\chi_j(2,3,1) + (-1)^{S_i+S_j} \chi_j(1,2,3)\chi_j(3,1,2) .$$

The wave functions of the valence shell are

$$\Psi_i(1,2) = \frac{1}{(2)^{1/2}} \left[\Psi_i(1)\Psi_i(2) + (-1)^{S_i} \Psi_i(2)\Psi_i(1) \right] , \qquad (7.15)$$

where the single-electron wave functions Ψ_i satisfy the Hartree-Fock equation. The way in which the radial equations are obtained and numerically solved is similar to that described in previous sections.

The calculation of the cross sections for electron scattering by alkaline-earth atoms is different in one aspect from that for alkali atoms: more complicated matrix elements arise in the exchange terms due to the inter-electron interaction. This problem has been analyzed in detail in [7.186].

We begin with the shape resonances observed in elastic scattering. Experimental studies of the elastic scattering of electrons by group II atoms in the near-threshold energy range have revealed features in the cross sections which are similar to those observed for alkali atoms and are, naturally, identified as being shape resonances. Transmission experiments revealed the ^2P shape resonances in the elastic scattering of electrons by Mg, Cd, Zn, and Hg atoms [7.187]. In more recent crossed beam experiments where the scattered electrons were detected, the total cross sections for electron scattering by Mg, Ca, Sr and

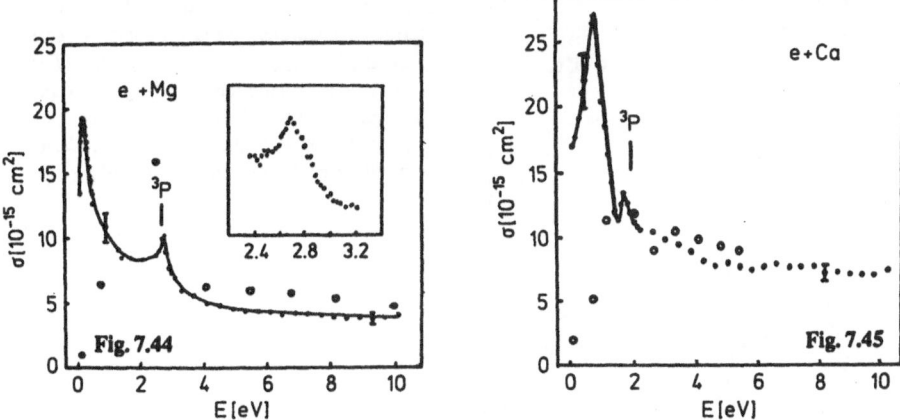

Fig. 7.44. Total cross section of e + Mg scattering. (•) experiment [7.190], (o) theoretical results [7.186]

Fig. 7.45. Total cross section of e + Ca scattering. (•) experiment [7.190], (o) theoretical results [7.186]

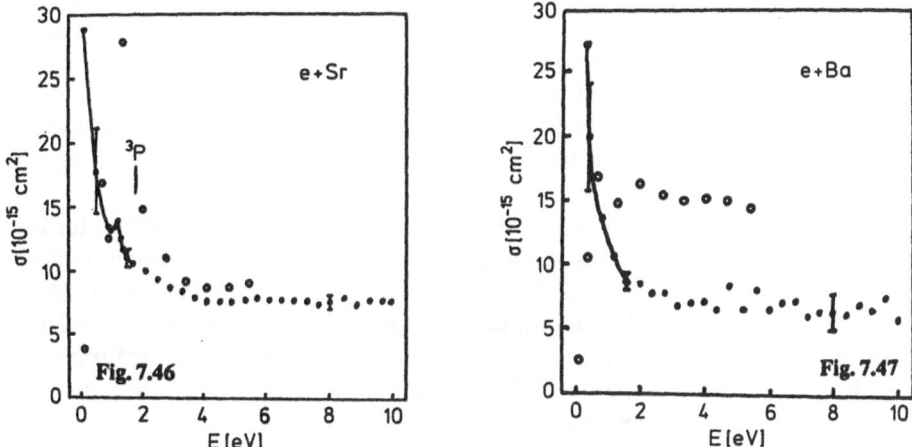

Fig. 7.46. Total cross section of e + Sr scattering. (•) experiment [7.190], (o) theoretical results [7.186]

Fig. 7.47. Total cross section of e + Ba scattering. (•) experiment [7.190], (o) theoretical results [7.186]

Ba atoms were measured over the energy interval of $0 - 10$ eV (Figs. 7.44–47) [7.188–190]. These experiments confirmed the existence of a ^2P resonance in e + Mg scattering and revealed the 2d shape resonances in the elastic scattering of electrons by Ca, Sr and Ba atoms.

The existence of the $^2P^0$ shape resonances for Be atoms has been predicted in several works [7.186, 191, 192], and also for Mg atoms [7.191–193]. Several authors [7.191, 192] solved the scattering problem in the one-channel approxi-

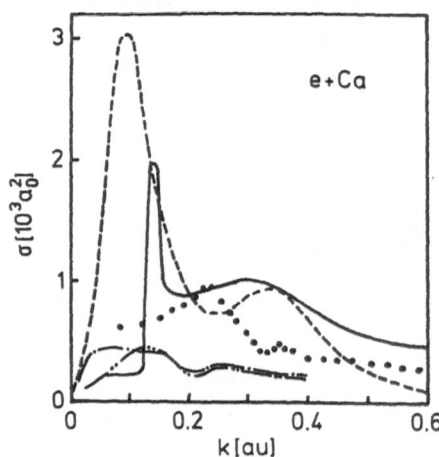

Fig. 7.48. Total cross section for the elastic scattering of electrons by the Ca atom. (•) experimental [7.188], (—) RPAE calculation [7.194], (— · —) CCM calculation including 4^1S, 4^1P, and 4^3P states of Ca [7.186]. (— ·· —) CCM calculation including 4^1S and 4^1P states of Ca, (- - -) simplified RPAE calculation [7.194]

mation with a potential $V = V_s + V_{pol}$ where V_s is the static potential of the atom, while V_{pol} is the phenomenological polarization potential. The form of the potential V_{pol} in [7.191] and [7.192] is different, and the energies of the resonances and their widths obtained in these works differ from each other considerably too. *Fabrikant* [7.186] and *Van Blerkom* [7.193] calculated the cross sections for elastic e + Be and e + Mg scattering using CCM for three states: n^1S, n^1P, and n^3P ($n = 3, 4$ for Be and Mg, respectively). The ^2P resonance was found in Fabrikant's calculations only for e + Be scattering. A possible explanation for the absence of a resonance for the Mg atom is an insufficiently small energy step.

CCM calculations have been carried out on elastic e + Ca scattering for two and three states [7.186] and in the RPAE [7.194]. The RPAE results indicate the existence of two maxima (Fig. 7.48) which the authors interpret as shape resonances. The first maximum formed by a p-wave at an energy of 0.27 eV was not found in the experiments [7.188]. The second maximum is formed primarily by the d-wave and lies at ~ 1.25 eV. The experiments of [7.188] revealed two 2d resonances in the elastic cross section, but the first resonance, a shape resonance, occurred at a far lower energy, ~ 0.7 eV, while the second which appeared at ~ 1.7 eV was classified by the authors as a Feshbach resonance. In the calculations [7.186], on the other hand, there was no P shape resonance at all near the elastic threshold, but at 1.77 eV the ^2D resonance of Feshbach type was revealed (Table 7.15). *Amusia* et al. [7.194] explained this by arguing that the calculations of *Fabrikant* [7.186] were carried out with crude semi-phenomenological wave functions for the Ca atom.

Since the two valence electrons of the group II atoms form a filled ns^2 shell, it had been assumed that these atom do not form stable negative ions until quite recently. However, in 1987 the existence of a stable $4s^2 4p$ ^2P state of the Ca$^-$ ion with a binding energy of $E_c = 0.045$ eV was established both theoretically [7.195] and experimentally [7.196]. Therefore, it was suspected that the ^2P shape resonance predicted theoretically is in reality caused by incomplete inclusion of

Table 7.15. Experimental energies [eV] and widths [eV] of shape resonances in elastic scattering of electrons by group II atoms, obtained in transmission experiments [7.187]

Atom	E	Γ
Mg	0.15	0.14
Cd	0.33	0.33
Zn	0.49	0.45
Hg	0.63	0.4

the polarizability of the Ca atom. If one does this correctly, this ^2P state will be shifted to below the ground state of Ca, while the energy of the state which yields the shape resonance must lie above the energy of the ground state. Indeed, in [7.197] in which the polarization potential has been taken into account, the p phase shift does not show resonance behaviour.

To identify the origin of the singularity at 0.7 eV in [7.189] was the cross section calculated for the elastic scattering of electrons by the Ca atom using a model with a polarization potential. The value of the parameter ϱ (see (7.10)) was chosen to ensure the existence of a stable Ca$^-$ state. The effective potential for $l = 2$ obtained is shown in Fig. 7.49. The explicit form is typical for a potential with a barrier which leads to a shape resonance (Chap. 2). It has been shown that there is no resonance in the p-wave while in the d-wave a shape resonance appears at 0.9 eV with a width of 0.68 eV.

Fabrikant [7.186] also studied e + Sr and e + Ba scattering using CCM with two and three states. These calculations revealed a ^2D shape resonance in elastic e + Sr scattering which had also been observed in experiment [7.188]. However, the calculations did not show the ^2D shape resonance in elastic e + Ba scattering which had also been observed [7.188]. A possible reason is that, as in the case of e + Mg scattering, the step along the energy scale was insufficiently small. Table 7.15 also shows the positions and widths of the shape resonances in the elastic scattering of electrons by these group II atoms.

We turn now to the Feshbach resonances which have been observed in collisions of electrons with other group II atoms.

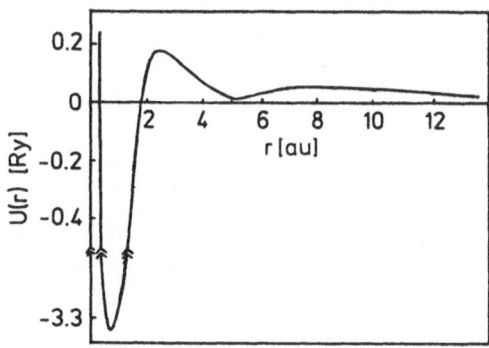

Fig. 7.49. A realistic potential V which provides the existence of D shape resonance in e + Ca elastic scattering

A resonance below the excitation threshold of the 3^3P level of the Mg atom, at an energy of 2.7 eV, was first observed in the experiment reported in [7.198]. *Burrow* and *Comer* believed that this resonance was due to the $3s3p^2\ {}^2D$ state of the Mg^- ion and they linked it to a D resonance which had been calculated theoretically in [7.186] to be at 2.6 eV. The same resonance has been experimentally observed in [7.189] and [7.190], which we have already mentioned, at 2.7 eV (Fig. 7.44). *Romanyuk* et al. [7.188] also observed resonances below the first excitation threshold in electron scattering by Ca and Sr atoms (Figs 7.45, 46). Although no resonances have been observed experimentally under the first excitation threshold of the Ba atom, the calculations [7.186] clearly reveal a rather broad maximum in the partial cross section $\sigma(L = 1)$ at 2.04 eV.

The study of resonances in the excitation cross sections of Zn and Cd atoms, for the 5^3S_0 and 6^3S_1 states, respectively was carried out by optical methods [7.199–202]. The optical excitation functions were mesured with a very small energy spread, $\Delta E \approx 80$ meV. Several features were observed. After the contribution of cascade transitions was separated out, it was possible to interpret these features as Feshbach resonances [7.200]. Figure 7.50 shows an example of the excitation functions for Cd $(5^3P_1 - 6^3P)$ and Cd $(5^1S_0 - 5^1P_1)$ lines [7.200].

We see that a common feature at about 7.25 eV is observed in both curves at the same energy. This confirms the multichannel nature of the decay of those AIS which cause these features.

A high-resolution study of the influence of the AIS on the optical excitation functions for higher n^1S-states ($n < 11$) of the Cd atom has been carried out in [7.202]. At energies above the ionization threshold all the excitation functions reveal maxima due to the same AIS. It is interesting that the positions of maxima with higher n are slightly shifted to the higher energy region. This effect is caused by the so-called post-collision interaction (PCI) which has been suggested by

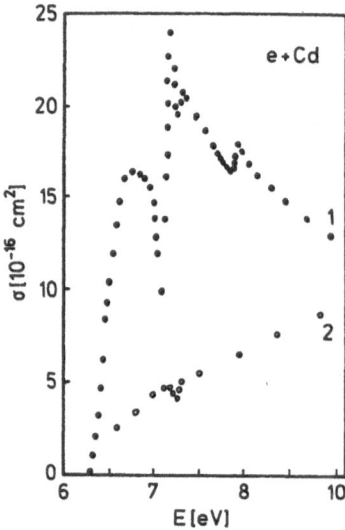

Fig. 7.50. Energy dependence of the excitation cross sections of lines of the Cd atom [7.200]. $1 - \lambda = 508.6$ nm $(5^3P_1 - 6^3S_1)$, $2 - \lambda = 228.8$ nm $(5^2S_0 - 5^1P_1)$

Hicks et al. [7.203] in electron-atom collisions. The PCI excitation mechanism which produces these maxima can be summarized by the reaction

$$e + Cd\,(5^1S_0) \rightarrow Cd^{**} + e_s \rightarrow Cd^+ + e_s + e_f \rightarrow Cd^*\,(n^1S_0) + e_f,$$

where Cd^{**} denotes the AIS and the subscripts s and f refer to the slow scattered and the fast ejected electrons, respectively.

Table 7.16 summarizes the experimental and theoretical results on the positions and classification of the resonances in elastic scattering cross sections (except shape resonances) and in the excitation cross sections for Ca, Sr, Ba, Mg, Zn, and Cd atoms.

Table 7.16. Energies [eV] and classification of resonances in elastic scattering cross sections and excitation cross sections of group II atoms

Atom	Experiment [7.188, 190, 204] a	Theory [7.186, 193] b	Configuration, term
Ca	1.7 ± 0.1	1.77	$3d^24s$ or $3d4s4p$ P
Sr	1.2 ± 0.1	0.95	$4d^25s$ or $4d5s5p$ D
Ba	3.92	–	$5s5p6s$ P
Mg	2.7 ± 0.3	2.58	$3s3p^2$ D
	3.1	–	$3s^23d$ ^2D
	4.4	–	$3s^24p$ $^2P^0$
	4.7	5.78	$3s3p^2$ ^2S
	5.2	–	$3s3p^2$ ^2P
	5.7	–	$3s3p4s$ $^4P^0$
	6.3	–	$3s3p3d$ $^2D^0$
	6.5	–	$3s3p3d$ $^2P^0$
	6.7	–	$3s3p5s$ $^2P^0$
	7.0	–	$3s3p4d$ $^2P_0$
	7.4	–	$3s3p5d$ $^2P^0$
Zn	7.18	–	–
	7.56	–	–
Cd	6.75	–	–
	7.24		

a – optical method; b – CCM

Resonances in the excitation cross sections of Hg atoms have also been studied in detail. The first hints of their existence appeared in investigations by *Shpenik* and *Zapesochny* [7.204] as early as 1965. The same year, *Kuyatt* and coworkers reeported the observation of 13 features in the energy dependence of an electron current passed through mercury vapour [7.205]. An analysis carried out by *Fano* and *Cooper* [7.206] made it possible to assign some of these features to the formation of short-lived states of the Hg^- ion. Figure 7.51 shows some excitation functions for Hg lines obtained in [7.199]. *Ottley* and *Kleinpoppen* [7.207] subsequently measured the optical excitation function of an intercombination line

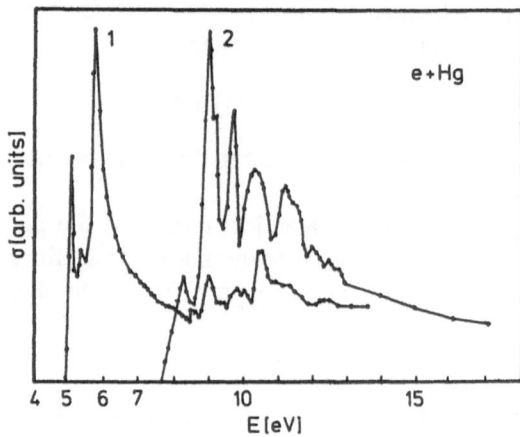

Fig. 7.51. Energy dependence of the excitation cross sections of lines of the Hg atom [7.199]. $1 - \lambda = 253.7$ nm ($6^1S_0 - 6^3P_1$), $2 - \lambda = 546.1$ nm ($6^3P_2 - 7^3S_1$)

of mercury and its polarization. The results of this paper agree well with those of a high resolution (~ 0.05 eV) study by *Koch* et al. [7.208] of optical excitation functions. Eleven resonances were found between 4.5 and 11.0 eV in [7.208].

Resonances in e + Hg scattering have since then been the subject of many studies. *Newman* et al. [7.209] measured the excitation cross section of metastable levels of Hg with an energy resolution of 25 meV. They observed a rich resonance structure between 8.5 and 11 eV.

Albert et al. [7.210] have measured the differential cross sections for elastic scattering over the interval 4.5 – 6 eV (Fig. 7.52). These measurements and also the theoretical analysis of all the results on resonances carried out by *Heddle* [7.211] provide a reliable classification of the resonances in this energy region. *Kazakov* et al. [7.212] have also measured the differential cross sections for elastic and inelastic scattering of electrons by Hg atoms through an angle of 90°. Their results on the elastic scattering support the data found in earlier studies.

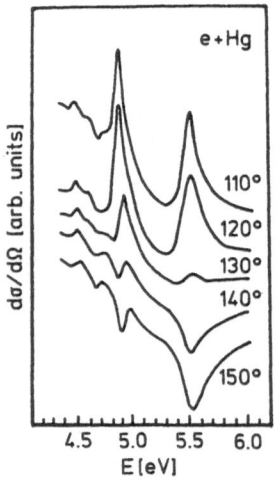

Fig. 7.52. Energy dependence of the differential cross sections for elastic scattering of electrons by the Hg atom [7.211]

Table 7.17. Energies [eV] and classification of resonances in e + Hg scattering

Classification [7.97, 209]	[7.37]	[7.187]	[7.208]	[7.207]	[7.211]	Theory [7.213]a
$6s^2 6p\ ^2P$	–	0.63	–	–	–	–
$6s6p^2\ ^4P_{1/2}$	4.07	–	–	–	4.55	4.7
$6s6p^2\ ^4P_{3/2}$	4.30	–	4.70	–	4.71	4.7
$6s6p^2\ ^4P_{5/2}$	4.89	–	4.90	4.92	4.94	4.9
$6s6p^2\ ^2D_{3/2}$	–	–	–	5.23	–	5.5
$6s6p^2\ ^2D_{5/2}$	–	–	–	5.50	5.51	5.5

a – R-matrix

The most comprehensive theoretical interpretation of elastic e + Hg scattering and of the excitation of the $6s6p\ ^3P_{0,1,2}$ and $6s6p\ ^1P_1$ states was provided by *Burke* et al. [7.213] who used an R-matrix method and considered five states at energies near the threshold. As can be seen from Table 7.17, the theoretical results agree nicely with the experimental data.

7.8 Polarization Effects in Scattering

Polarization experiments can reveal important details of atomic collision processes. Progress in the development of sources of polarized electrons and of intense atomic beams polarized by laser radiation has provided a new impetus to research on spin-dependent effects in electron-atom collisions.

We recall that the primary quantities are amplitudes which depend, at a given energy, on the projections of the spins of the incident electron and of the atomic target in the initial and final states. The total cross section is integrated over angles, summed over the projections of the final-state spins, and averaged over the projections of the initial-state spins. Consequently, measurements of total cross sections alone do not provide complete information on the amplitude. A "complete" experiment allows one to determine the magnitude and phase of the amplitude for given values of the spin variables. It provides the maximum amount of information about the process and allows the most comprehensive comparison of theory and experiment. In particular, such an experiment makes it possible to determine separately the cross sections for direct and exchange scattering.

As we have already pointed out, coincidence experiments can be used to determine the magnitudes and relative phase shifts of the amplitudes for the excitation of an atom into various magnetic substates of a given state. Experiments with polarized beams provide information on the spin dependence of electron-atom collisions. A coincidence experiment with polarized beams would, in principle, allow one to determine the scattering amplitude [7.214]. General questions regarding the theory of polarization effects in electron-atom collisions are discussed in [7.215–217]. A general overview of experiments on polarized-electron scattering was given by *Kessler* [7.218], and *Drukarev* and *Obyedkov* [7.219] where the related literature is cited abundantly.

In polarization experiments, a polarized or unpolarized electron beam is scattered by unpolarized or polarized atoms. Polarized beams of electrons and atoms can be produced in various ways: "Chemionization" of optically oriented He atoms in the 2^3S state, via the Fano effect in the scattering of electrons by alkali atoms, and the photoionization of polarized atoms, among others, see, e.g., [7.220, 221] and detailed lists of references therein. A popular method is to use photoemission from a GaAs cathode illuminated by circularly polarized laser light [7.221]. *Pierce* et al. [7.221] produced an electron beam with a 43 % polarization at a current of 20 μA with an energy spread $\Delta E = 150$ meV. Polarized electrons have also been produced by scattering of a beam of unpolarized electrons by a target of some sort ("double-scattering experiments" [7.222]) A polarized atomic beam can be produced by sending an unpolarized atomic beam through a nonuniform magnetic field filter (a Stern-Gerlach experiment). Another technique widely used today is to optically pump a substate of the hyperfine structure of an atom with circularly polarized laser light. This technique achieves an 85 % polarization of an atomic beam.

There are two types of polarization experiments, differing in the quantities which are measured. In the first type, one measures cross sections for processes with given spin states of the particles after scattering or some other quantities which characterize the spin states of the electrons and atoms after scattering. In the second type, one measures the polarization of the photons which are emitted by atoms excited in a collision process.

Among the experiments of the first type are those on the excitation of polarized alkali metal atoms by polarized electrons done by *Baum* et al.[7.223, 224]. They measured an asymmetry parameter A_{ns-np} defined as

$$A_{ns-np} = \frac{\sigma(\uparrow\downarrow) - \sigma(\uparrow\uparrow)}{\sigma(\uparrow\downarrow) + \sigma(\uparrow\uparrow)} . \tag{7.16}$$

Here $\sigma(\uparrow\uparrow)$ and $\sigma(\uparrow\downarrow)$ are the excitation cross sections for the cases in which the spins of the electron and atom are, respectively, parallel and antiparallel.

For the parallel spin configuration of the initial state of the particles, the process is described by a triplet cross section σ_T. For an antiparallel spin configuration, which is a mixture of singlet and triplet states, the cross section is

$$\tfrac{1}{2}(\sigma_T + \sigma_S) . \tag{7.17}$$

In terms of the singlet and triplet cross sections, the asymmetry can thus be written

$$A_{ns-np} = \frac{\sigma_S - \sigma_T}{4\sigma} , \tag{7.18}$$

where σ is the total cross section given by

$$\sigma = \tfrac{1}{4}\sigma_S + \tfrac{3}{4}\sigma_T .$$

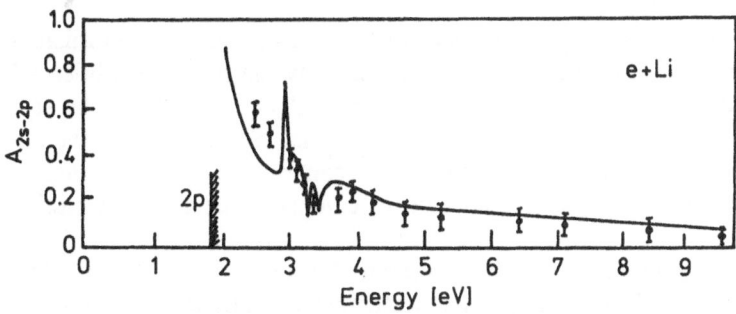

Fig. 7.53. The asymmetry parameter A_{2s-2p} for e + Li scattering. *Points* experiment [7.225], *solid line* DM calculation [7.226]

Figure 7.53 shows results which have been obtained on the asymmetry parameter A_{2s-2p} for the Li atom. This experiment was carried out by *Baum* et al. [7.225]. Theoretical results have been obtained by the DM [7.226] and by the CCM [7.227]. It can be seen from this figure that the calculations of the asymmetry in the DM indicate the presence of the features due to the ^{1}S, ^{3}P, and ^{1}D resonances (Figs. 7.18, 19). On the other hand, no resonances have been discovered experimentally. To discover them one requires a smaller energy step between data points. As follows from (7.18), the background substraction in σ_S and σ_T enables one to see the resonances in the asymmetry parameter more clearly than in the nonpolarized electron scattering experiments.

Figure 7.54 shows the results of a theoretical calculation [7.138] of the polarization vector of the elastically scattered electrons in e + Cs (6s) collisions for various scattering angles and for various polarization states of the beams. If the beam is incident along the Z axis, and if the scattering plane is the (X, Z) plane,

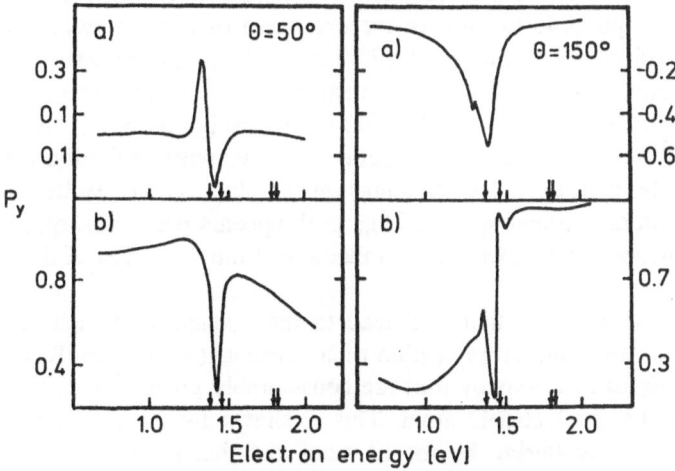

Fig. 7.54. The polarization vector P_y in e + Cs scattering for various scattering angles and for (a) an unpolarized electron beam, (b) an electron beam polarized in the transverse direction. From [7.138]

the polarization vector P is perpendicular to this plane, and the nonvanishing component $P_y(|P_y| \leq 1)$ is equal to the difference between the probabilities W_1 and W_2 for electron spin projections $S_y = 1/2$ and $S_y = -1/2$. It can be seen from these figures that P_y is small outside the resonance region, while in the resonance region the curve depends strongly on the scattering angle. Unfortunately, we do not yet have any experimental data to compare with these theoretical results.

Such data are available for e + Hg scattering, however. *Albert* et al. [7.211] measured the spin polarization of electrons scattered elastically by mercury atoms as a function of the energy over the interval 4–6 eV at several angles. They determined the positions and classification of three resonances: $^2S_{1/2}$ (4.55 eV), $^2D_{3/2}$ (4.71 eV) and $^2D_{5/2}$ (4.49 eV).

Although experiments of the second type have been done a long time ago, with electrons and atoms initially unpolarized, it is only in the last decade that such experiments have been carried out with polarized beams of atoms and electrons. *Jitschin* et al. [7.228] measured the degree of polarization of the radiation from polarized Na atoms excited by an unpolarized electron beam. No special study was made of resonances in those experiments, however, and there is no evidence of them on the curve of the polarization as a function of the energy of the incident electrons. An experiment to measure the circular polarization of the radiation emitted after the excitation of the 6s6p ^2P level of the mercury atom was used in [7.229] to classify resonances.

Although so far there have been relatively few studies of resonances in polarization experiments, the rapid progress which we have recently seen in improvement of the apparatus for producing polarized atoms and electrons raises the hope that many interesting results will be obtained in the near future.

The collisions of electrons with excited atoms is an another prospective direction in resonance studies. We have considered up to now only experiments where the initial state of the atom is the ground state. Improvements in experimental apparatus have made it possible to observe the excitation of excited atoms by electrons (stepwise excitation). *Stumpt* et al. [7.3] not so long ago observed the excitation of the Na atom from the 3p level. An extensive research program has been carried out by *Aleksakhin* et al. [7.230, 231] on the excitation of Tl^+, Na^+, Cu^+, Ba^+, and Sr^+. Interestingly, the cross sections for stepwise excitation are significantly larger – by a factor of 10^2 in some cases – than for the excitation of the same levels from the ground state. A large peak appears near threshold; it may be a shape resonance. As yet, we have no theoretical interpretation of these results.

In conclusion, we may assert that AIS lead to the resonances which can be several times higher than the cross section of nonresonant processes. Resonances in the scattering of electrons by ions are considerably greater than those for the scattering by the isoelectronic atom. This is shown by the differential cross section for electron scattering by the Li atom and that for the Be^+ ion (Figs. 7.21, 37) or by the total excitation cross section for the 3p level of Na and that for the Mg^+ ion (Figs. 7.28, 42). The higher the charge of the ion, the larger is the resonance contribution. This trend is illustrated by the electron scattering by the Be-like ions [7.232]. In this work, it was shown that resonance might enhance the electron-impact excitation cross section by a factor of 2 – 4 for the electric-dipole forbidden transitions while for the dipole-allowed ones the resonance effects amount to only a few per cent of these rates. It should be noted, however, that from some highly charged ions the resonance effects become less significant because of increasing dominance of the radiative channel of the AIS decay [7.233].

For some multiply charged ions (e.g., Se^{24+}), the resonance contribution, even to averaged quantities such as the excitation rate coefficients in plasma at low (0 – 2 eV) temperatures, may reach a factor of 4 [7.234]. Therefore, there is a need to make further experimental tests of the effect of formation of dielectronic resonances. As we have seen, theoretically these resonances can influence the average cross sections by substantial factors, and microscopic cross sections are affected even more.

8. Conclusion

The material presented in this book shows that resonances in the cross sections for scattering of electrons by atoms and ions are now an important and urgent problem both theoretically and experimentally. The influence of resonances on the cross section increases with increasing charge of the ion. As *Smirnov* [8.1] has pointed out, the presence of AIS of multiply charged ions has a strong effect on the nature of the emission spectrum of a hot plasma. The formation and decay of AIS strongly influence the kinetics of the filling of levels in a low-temperature laser plasma. The role played by AIS in plasma was studied in detail in [8.2].

The number of atoms for which we have a comprehensive picture of the resonances is not much greater than 10. So far, only the first pages have been written on the spectroscopy of AIS of atoms and ions. Resonance phenomena are a touchstone for testing theoretical models for describing multiparticle interactions. The methods available today, such as the method of the R-matrix, the CCM and the DM have been fairly successful in describing resonance effects at low energies. Despite this progress, however, the role played by electron-electron correlations in the formation of AIS has not been studied adequately. As this book has shown, most of the work, both experimental and theoretical, has dealt with collisions of electrons with light atoms. We can apparently expect to see a greater effort on detailed studies of heavy atoms in the future. This problem seems to be completely solvable in view of the significant recent progress in relativistic atomic theory.

Another factor which makes this an important problem is that the topic of resonance phenomena goes beyond the scope of electron-atom collisions, also including resonances in the electron scattering by molecules, in atom collisions, in nuclear scattering, and in the scattering of elementary particles. We have not taken up here such interesting and important processes in which resonances are manifested as the recombination of ions, electron-impact ionization, single- and multiphoton ionization of atoms and molecules, charge transfer, and the resonances in electron scattering by atoms and molecules in a strong laser field. Each of these topics has been the subject of a formidable list of studies and of special monographs [8.3–6].

When the surface of a metal is bombarded with positive ions, negative ions also form on the surface. There is the hope that an expansion of this research will lead to new directions in research on condensed matter.

We might point out yet another area in which AIS play an important role, namely, relativistic nuclear physics which requires understanding the ionization of atoms and ions by relativistic electrons [8.7]. At electron energies of a few MeV, AIS are excited due to the formation of vacancies in inner shells. Theoretically, the problem of the scattering of relativistic electrons requires solution of the relativistic analog of the Schrödinger equation: the Dirac equation. Such calculations have not yet been carried out comprehensively. These processes also pertain to so-called processes with a redistribution of particles. As yet, we do not have an adequate theory for such processes. Much hope here is pinned on an approach based on the use of Faddeev equations [8.8].

Important success has been recently achieved in microscpic description of resonances in nuclear processes. This problem has been tackled by using so-called resonance groups method which is, basically, an exact nuclear version of the CCM.

Since in all these cases we are dealing with a many-body problem, we are tempted to work towards the construction of a unified theory for describing resonances in collisions of composite systems. Electron-atom collisions which we have discussed here may serve as a proving ground for a unified theory of this sort, since the potential of the interaction between particles is known. This goal can only be reached, however, after more experimental information has been acquired. Here it is very important to obtain more detailed total cross sections in experiments with beams with very high energy resolution and also differential cross sections, to carry out phase-shift analysis, and also to carry out experiments with polarized beams.

References

Chapter 1

1.1 N.F. Mott, H.S.W. Massey: *Theory of Atomic Collisions* (Clarendon, Oxford 1965)
1.2 P.G. Burke: *Potential Scattering in Atomic Physics* (Plenum, New York 1977)
1.3 H.S.W. Massey: *Atomic and Molecular Collisions* (Taylor and Francis, London 1979)
1.4 H.S.W. Massey: *Negative Ions* (Cambridge Univ. Press, Cambridge 1976)
1.5 R.K. Nesbet: *Variational Methods in Electron-Atom Scattering Theory* (Plenum, New York 1980)
1.6 G.F. Drukarev: *Collisions of Electrons with Atoms and Molecules* (Nauka, Moscow 1978) (in Russian)
1.7 V.P. Zhigunov, B.N. Zakhariev: *The Close Coupling Method in Scattering Theory* (Atomizdat, Moscow 1974) (in Russian)
1.8 A.G. Sitenko: *Scattering Theory*, Springer Ser. Nucl. Part. Phys. (Springer, Berlin, Heidelberg 1990)
1.9 B.N. Zakhariev, A.A. Suzko: *Direct and Inverse Problems. Potentials in Quantum Scattering* (Springer, Berlin, Heidelberg 1990)
1.10 V.V. Balashov: "The Contemporary State of Resonance Theory in Atomic Systems", in *Lecture Notes of the 1st School on Electronic and Ionic Collisions* (Phys. Techn. Inst. Ed., Kharkov 1969) p. 1 (in Russian)
1.11 V.I. Lengyel, M. Salak: *Nonrelativistic Scattering Theory* (Vyshcha Shkola, Lvov 1983) (in Russian)
1.12 B.L. Moiseiwitch: "Atomic Collision Theory", in *Atoms, Molecules and Lasers*, (Lectures Presented at the International Winter College in Trieste 17 January – 10 April 1973), (At. Energy Agency, Vienna 1974) p. 471
1.13 W. Eissner, P. Hummer, I. Percival (eds.): *Atoms in Astrophysics* (Plenum, New York 1983)
1.14 F. Brouillard, J.W. McGowan (eds.): *Physics of Ion-Ion and Electron-Ion Collisions* (Plenum, New York 1983)
1.15 G.J. Schulz: Rev. Mod. Phys. **45**, 378 (1973)
1.16 K. Smith: Repts. Progr. Phys. **29**, 373 (1966)
1.17 P.G. Burke: Adv. Atom. Molec. Phys. **4**, 173 (1968)
1.18 J. Callaway: Comments Atom. Molec. Phys. **10**, 279 (1981)
1.19 D.E. Golden: Adv. Atom. Molec. Phys. **14**, 1 (1978)
1.20 R. Whiddington, H. Priestley: Proc. Roy. Soc. A **145**, 462 (1934)
1.21 U. Fano: Phys. Rev. A **124**, 1866 (1961)
1.22 P.G. Burke, H. Shey: Phys. Rev. A **126**, 147 (1962)
1.23 M. Gailitis, R. Damburg: Proc. Phys. Soc. **82**, 192 (1963)
1.24 G.J. Schulz: Phys. Rev. Lett. **10**, 104 (1963)
1.25 G.J. Schulz: Phys. Rev. A **136**, 650 (1964)
1.26 C.E. Kuyatt, J.A. Simpson, S.R. Mielczarek: Phys. Rev. A **138**, 385 (1965)
1.27 I.P. Zapesochny, O.B. Shpenik: Zh. Exp. Teor. Fiz. **50**, 890 (1966)/English transl.: Sov. Phys. JETP **23**, 592 (1966)
1.28 D. Andrick, M.Eyb, M. Hofmann: J. Phys. B **5**, L15 (1972)
1.29 D. Burrow, J. Comer: J. Phys. B **8**, L92 (1975)

1.30 N.I. Romanyuk, O.B. Shpenik, I.P. Zapesochny: Zh. Exp. Teor. Fiz. Pis'ma **32**, 472 (1980)/English transl.: Sov. Phys. JETP Lett. **32**, 452 (1980)

1.31 O.B. Shpenik, L. Szóvtér, A.N. Zavilopulo, I.P. Zapesochny, E.E. Kontrosh: Zh. Exp. Teor. Fiz. **69**, 48 (1975)/English transl.: Sov. Phys. JETP **42**, 23 (1975)

1.32 J.N. Bardsley, F. Mandl: Repts. Progr. Phys. **31**, 472 (1968)

Chapter 2

2.1 N.F. Mott, H.S.W. Massey: *Theory of Atomic Collisions* (Clarendon, Oxford 1965)

2.2 M.A. Crees, M.J. Seaton, P.M.H. Wilson: Comput. Phys. Commun. **15**, 23 (1978)

2.3 R. Newton: *Scattering Theory of Waves and Particles* (McGraw-Hill, New York 1967)

2.4 V. de Alfaro, T. Regge: Potential Scattering (North-Holland, Amsterdam 1965)

2.5 J.R. Taylor: *Scattering Theory on Nonrelativistic Collisions* (John Wiley & Sons Inc., New York 1972)

2.6 T.F. O'Malley, L. Spruch, L. Rosenberg: J. Math. Phys. **2**, 491 (1961)

2.7 O. Dumbrajs, M. Martinis: J. Phys. B **15**, 961 (1981)

Chapter 3

3.1 V.P. Zhigunov, B.N. Zakhariev: *The Close-Coupling Method in Scattering Theory* (Atomizdat, Moscow 1974) (in Russian)

3.2 V.I. Lengyel, M. Salak: *Nonrelativistic Scattering Theory* (Vyshcha Shkola, Lvov 1983) (in Russian)

3.3 L. Castillejo, I.C. Percival, M.J. Seaton: Proc. Roy. Soc. **254**, 259 (1959)

3.4 N.F. Mott, H.S. Massey: *Theory of Atomic Collision* (Clarendon, Oxford 1965)

3.5 G.F. Drukarev: *Collisions of Electrons with Atoms and Molecules* (Nauka, Moscow 1978) (in. Russian)

3.6 M.Ya. Amusia, N.A. Cherepkov, S.G. Shapiro: Zh. Exp. Teor. Fiz. **63**, 889 (1972)

3.7 B.L. Moiseiwitch: "Atomic Collision Theory", in *Atoms, Molecules and Lasers* (Lectures Presented at the International Winter College in Trieste 17 January - 10 April 1973), (At. Energy Agency, Vienna 1974) p. 471

3.8 Y. Hahn: Phys. Rev. A **16**, 1964 (1977)

3.9 I.C. Percival, M.J. Seaton: Proc. Camb. Phil. Soc. **53**, 654 (1975)

3.10 P.G. Burke, K. Smith: Rev. Mod. Phys. **34**, 458 (1962)

3.11 K Smith: Repts. Progr. Phys. **29**, 373 (1966)

Chapter 4

4.1 H. Feshbach: Ann Phys. **5**, 357 (1958)

4.2 M. Goldberger, K.M. Watson: *Collision Theory* (John Wiley & Sons, Inc., New York 1964)

4.3 U. Fano: Repts. Progr. Phys. **46**, 97 (1983)

4.4 M. Gailitis, R. Damburg: Proc. Phys. Soc. **82**, 192 (1963)

4.5 R. Newton: *Scattering Theory of Waves and Particles* (McGraw-Hill, New York 1967)

4.6 P.G. Burke: Adv. Atom. Molec. Phys. **4**, 173 (1968)

4.7 G. Shulz: Phys. Rev. Lett. **10**, 104 (1963)

4.8 J.W. McGowan, E.M. Clarke, E.K. Curly: Phys. Rev. Lett. **15**, 917 (1965)

4.9 P.G. Burke, A.J. Taylor: Proc. Phys. Soc. **88**, 549 (1966)

4.10 W. Koshmider, V. Raible, H. Kleinpoppen: Phys. Rev. A **128**, 1365 (1973)

4.11 U. Fano: Phys. Rev. A **140**, 67 (1965)

4.12 K. Smith, L.A. Morgan: Phys. Ref. A **165**, 110 (1968)

4.13 V.I. Lengyel, V.T. Navrotsky, E.P. Sabad: *Theory of Resonance Phenomena in Electron-Atom Scattering* (Naukova Dumka, Kiev 1988) (in Russian)

Chapter 5

5.1 V.V. Balashov, S.I. Grishanova, I.M. Kruglova, V.S. Senashenko: Phys. Lett. **27A**, 101 (1968)
5.2 V.V. Balashov, S.I. Grishanova, I.M. Kruglova, V.S. Senashenko: Opt. Spektrosk. **28**, 859 (1970)/English transl.: Opt. Spectrosc. **28**, 466 (1970)
5.3 V.V. Balashov, S.S. Lipovetsky, A.V. Pavlichenkov, A.N. Poljudov, V.S. Senashenko: Vestnik of Moscow University, ser. Phys. Astron. **12**, 65 (1971) (in Russian)
5.4 H. Feshbach: Ann. Phys. **5**, 357 (1958)
5.5 M.I. Gaysak, V.I. Lengyel, V.T. Navrotsky, E.P. Sabad: Ukr. Fiz. Zh. **25**, 1329 (1980)
5.6 M.K. Gailitis: Zh. Exp. Teor. Fiz. **47**, 160 (1964)
5.7 P.G. Burke: Adv. Atom. Molec. Phys. **4**, 173 (1968)
5.8 A.I. Baz', Ya.B. Zel'dovich, A.M. Perelomov: *Scattering, Reactions and Decay in Non-Relativistic Quantum Mechanics* (Nauka, Moscow 1971) (in Russian)
5.9 V.P. Zhigunov, B.N. Zakhariev: *Close-Coupling Method in the Scattering Theory* (Atomizdat, Moscow 1974) (in Russian)
5.10 N.F. Mott, H.S.W. Massey: *Theory of Atomic Collisions* (Clarendon, Oxford 1965)
5.11 M.E. Rudd: Phys. Rev. Lett. **13**, 503 (1964)
5.12 D.R. Herrick, O. Sinagoglu: Phys. Rev. A **11**, 97 (1975)
5.13 P.G. Burke, D.D. McVicar: J. Phys. B. **86**, 989 (1965)
5.14 P.J. Hick, J. Comer: J. Phys. B **8**, 1866 (1975)
5.14 R.P. Madden, K. Codling: Phys. Rev. Lett. **10**, 516 (1963)
5.15 R.P. Madden, K. Codling: Astrophys. J. **141**, 364 (1965)
5.15 L. Lipsky, R. Anania, M.J. Connelly: *Atomic Data and Nuclear Data Tables* **20**, 127 (1977)
5.16 U. Fano, J.W. Cooper: Phys. Rev. A **137**, 1365 (1965)
5.17 G.F. Drukarev: *Theory of Collisions of Electrons with Atoms* (Fizmatgiz, Moscow 1963) (in Russian)
5.18 V.V. Balashov, A.N. Grum-Grzymailo, N.M. Kabachnik, A.I. Magunov, S.I. Strakhova: J. Phys. B **12**, 2233 (1979)
5.19 O.I. Zatsarinny, V.I. Lengyel, E.P. Sabad: Proc. XIII Int. Conf. Phys. Electron. Atom. Coll. (Berlin, 1983) p. 749

Chapter 6

6.1 D. Hartree: *Calculation of Atomic Structures* (Wiley, New York 1957)
6.2 C. Froese-Fischer: *The Hartree-Fock Method for Atoms. A Numerical Approach* (Wiley, New York 1977)
6.3 M.Ya. Amusia, N.A. Cherepkov: Case Studies Atom. Phys. **5**, 47 (1975)
6.4 A.V. Fock: Z. Phys. B **98**, 145 (1936)
6.5 J.H. Macek: J. Phys. B **1**, 831 (1968)
6.6 C.D. Lin: Adv. At. Mol. Phys. **22**, 77 (1986)
6.7 H. Clar, M. Clar: J. Phys. B **13**, 1057 (1980)
6.8 J.E. Hornos, S.W. McDowell, C.D. Coldwell: Phys. Rev. A **33**, 2212 (1986)
6.9 C.H. Green: Phys. Rev. A **26**, 2974 (1982)
6.10 A.G. Abrashkevich, M.I. Gaysak, V.I. Lengyel, V.Yu. Pojda, I.V. Puzynin: Phys. Lett. A **33**, 140 (1988)
6.11 H.G.M. Heideman: "The Formation and Decay of Triply Excited He⁻ States in e–He Scattering", in *Multiphoton Processes* (Proceedings of the 4th ICOMP, Boulder, Colorado July 13–17 1987) ed. by S.J. Smith, P.L. Knight (Cambridge Univ. Press, Cambridge 1988)
6.12 J.W. Cooper, U. Fano, F. Prats: Phys. Rev. Lett. **10**, 518 (1963)

6.13 U. Fano: Phys. Rev. A **140**, 67 (1965)

6.14 K. Smith, L.A. Morgan: Phys. Rev. A **165**, 110 (1968)

6.15 D.W. Norcross: J. Phys. B **2**, 1300 (1969)

6.16 V.I. Lengyel, V.T. Navrotsky, E.P. Sabad: *Theory of Resonance Phenomena in Electron-Atom Scattering* (Naukova Dumka, Kiev 1988) (in Russian)

6.17 P.G. Burke: Adv. Atom. Molec. Phys. **4**, 173 (1968)

6.18 P.G. Burke, M.J. Seaton: in *Methods in Computational Physics*, Vol. **10**, ed. by B. Adler, S. Fernbach, M. Rotenberg (Academic, New York 1971)

6.19 M. Gailitis, R. Damburg: Zh. Exp. Teor. Fiz. **44**, 1644 (1963)/English transl.: Sov. Phys. JETP. **17**, 1107 (1963

6.20 N.F. Mott, H.S.W. Massey: *Theory of Atomic Collisions* (Clarendon, Oxford 1965)

6.21 V.V. Balashov, S.I. Grishanova, I.M. Kruglova, V.S. Senashenko: Opt. Spektrosk. **28**, 859 (1970)/English transl: Opt. Spectrosc. USSR **28**, 859 (1970)

6.22 A.M. Lane, R.G. Thomas: Rev. Mod. Phys. **30**, 257 (1958)

6.23 I.C. Percival, M.J. Seaton: Proc. Camb. Phil. Soc. **53**, 654 (1957)

6.24 J. Hahn, T.F. O'Malley, L. Spruch: Phys. Rev. A **134**, 397 (1964)

6.25 M. Gailitis: Zh. Exp. Teor. Fiz. **47**, 160 (1964)

6.26 M.I. Gaysak, V.I. Lengyel, V.T. Navrotsky, E.P. Sabad: Ukr. Fiz. Zh. **25**, 1328 (1980)

6.27 W.A. Lester: J. Comput. Phys. **3**, 322 (1968)

6.28 M.J. Seaton: J. Phys. B **7**, 1817 (1974)

6.29 I.P. Zapesochny, A.I. Dashchenko, A.I. Imre, A.N. Gomonaj, V.I. Lengyel, V.T. Navrotsky, E.P. Sabad: Pis'ma Zh. Exp. Teor. Fiz. **39**, 45 (1984)/English transl.: Sov. Phys. JETP Lett. **39**, 45 (1984)

6.30 M.A. Crees, M.J. Seaton, P.M.H. Wilson: Comput Phys. Commun. **15**, 23 (1978)

6.31 E.P. Wigner, L. Eisenbud: Phys. Rev. **72**, 29 (1974)

6.32 P.G. Burke, A. Hibbert, D.W. Robb: J. Phys. B **4**, 153 (1971)

6.33 R.K. Nesbet: *Variational Methods in Electron-Atom Scattering Theory* (Plenum, New York 1980)

6.34 M.Ya. Amusia, A.S. Kheifets: Phys. Lett. A **82**, 407 (1981)

Chapter 7

7.1 L. Sanche, G.J. Schulz: Phys. Rev. A **5**, 1672 (1972)

7.2 M. Eyb, M. Hofmann: J. Phys. B **8**, 1095 (1975)

7.3 B. Stumpt, A. Gallagher: Proc. 13th Int. Conf. Phys. Electron. Atom Collisions, Post-Post Deadline Papers (Berlin, 1983) p. 3

7.4 F.H. Read: Fysica Scripta **27**, 103 (1983)

7.5 R.E. Kennerly, R.J. Van Brunt, A.C. Gallagher: Phys. Rev. A **23**, 2430 (1981)

7.6 V.I. Frontov: Opt. Spektrosk. **59**, 460 (1985)/English transl.: Opt. Spectrosc. USSR **59**, (1985)

7.7 V.I. Lengyel, V.T. Navrotsky, E.P. Sabad: *Theory of Resonance Phenomena in Electron-Atom Collisions* (Naukova Dumka, Kiev 1988) (in Russian)

7.8 O.B. Shpenik, A.N. Zavilopulo, A.V. Snegursky, I.I. Fabrikant: J. Phys. B **17**, 887 (1984)

7.9 I.I. Fabrikant, O.B. Shpenik, A.V. Snegursky, A.N. Zavilopulo: Phys. Repts. **159**, 1 (1988)

7.10 V.V. Balashov, S.S. Lipovetsky, V.S. Senashenko: Phys. Lett. A **38**, 103 (1972)

7.11 K. Smith, R.P. McEachran, P.A. Fraser: Phys. Rev. A **125**, 553 (1962)

7.12 P.G. Burke, H. Shey: Phys. Rev. A **126**, 147 (1962)

7.13 P.G. Burke, A.J. Taylor: Proc. Phys. Soc. **88**, 549 (1966)

7.14 M. Gailitis, R. Damburg: Proc. Phys. Soc. **82**, 526 (1963)

7.15 G.J. Schulz: Phys. Rev. Lett. **13**, 583 (1964)

7.16 H. Kleinpoppen, V. Raible: Phys. Lett. **18**, 24 (1965)

7.17 S. Ormonde, J. McEven, J.W. McGovan: Phys. Rev. A **22**, 1165 (1969)

7.18 J.W. McGovan, E.M. Clarke, E.K. Curly: Phys. Rev. Lett. **15**, 917 (1965)

7.19 C.D. Warner, G.C. King, P. Hammond, J. Slevin: Proc. 14th Int. Conf. Phys. Electron. Atom Coll. (Palo Alto, 1985) p. 142. See also: J. Phys. B **19**, 3297 (1986)

7.20 L.A. Morgan, M.R.C. McDowell, J. Callaway: J. Phys. B **10**, 3297 (1977)

7.21 J. Callaway: Phys. Rev. A **26**, 199 (1982)

7.22 F.H. Read: Aust. J. Phys. **35**, 475 (1982)

7.23 C.D. Lin: Phys. Rev. Lett. **51**, 1348 (1983)

7.24 A.R.P. Rao: J. Phys. B **16**, 699 (1983)

7.25 H. Fukuda, N. Kayama, M. Matsuzawa: J. Phys. B **20**, 2959 (1987)

7.26 J.F. Williams: J. Phys. B **9**, 1519 (1976); J. Phys. B **21**, 2107 (1988)

7.27 P.M. Rutter, C.D. Warner, G.C. King: Proc. 15th Int. Conf. Phys. Electron. Atom Coll. (Brighton, 1987) p. 221; J. Phys. B **23**, 93 (1990)

7.28 A. Kuppermann, D. Hood, T. Natson: Ibid., p. 193

7.29 J. Callaway: Phys. Rev. A **37**, 3699 (1988)

7.30 Y.K. Ho, J. Callaway: Phys. Rev. A **34**, 1307 (1986)

7.31 A. Pathak, A.E. Kingston, K.A. Berrington: J. Phys. B **21**, 2939 (1988)

7.32 C.A. Nicolaides, Y. Komninos: Phys. Rev. A **35**, 999 (1987)

7.33 Y.K. Ho: Phys. Rev. A **38**, 6424 (1988)

7.34 V. Melezhik, F. Vukajlović: Phys. Rev. A **38**, 6426 (1988)

7.35 J. Botero: Phys. Rev. A **35**, 36 (1987)

7.36 G.J. Schulz: Phys. Rev. Lett. **10**, 104 (1963)

7.37 C.E. Kuyatt, J.A. Simpson, S.R. Mielczarek: Rev. Sci. Instrum. **34**, 1454 (1963)

7.38 K.A. Berrington, P.G. Burke, A.L. Sinfailam: J. Phys. B **8**, 1459 (1975)

7.39 P.G. Burke, L.C.G. Freitas, A.E. Kingston, K.A. Berrington, A. Hibbert, A.L. Sinfailam: J. Phys. B **17**, L303 (1984)

7.40 B.F. Davos, K.T. Chung: Phys. Rev. A **29**, 1878 (1984)

7.41 D.E. Golden, F.D. Schowengerdt, J. Macek: J. Phys. B **7**, 478 (1974)

7.42 S. Cvejanović, J. Comer, F.H. Read: J. Phys. B **7**, 468 (1974)

7.43 J.N.H. Brunt, G.C. King, F.H.J. Read: J. Phys. B **10**, 1289 (1977)

7.44 D.E. Golden, A. Zecca: Rev. Sci. Instrum. **42**, 210 (1971)

7.45 D. Andrick, L. Laghans: J. Phys. B **8**, 1245 (1975)

7.46 D.W.O. Heddle: Contemp. Phys. **17**, 443 (1976)

7.47 D. Spence, D. Stuit, M.A. Dillon, R.G. Wang: Proc. 13th Int. Conf. Phys. Electron Atom. Coll. (Berlin, 1983) p. 117

7.48 R.S. Oberoj, R.K. Nesbet: Phys. Rev. A **8**, 2969 (1973)

7.49 P.G. Burke, L.C.G. Freitas, A.E. Kingston, K.A. Berrington: Proc. 16th Int. Conf. Phys. Electron. Atom. Coll. (Berlin, 1983) p. 108

7.50 S.J. Buckman, P. Hammond, F.H. Read, G.C. King: J. Phys. B **16**, 4039 (1983)

7.51 S.J. Buckman, D.N. Newman: J. Phys. B **20**, L711 (1987)

7.52 J. Jureta: Proc. 16th Int. Conf. Phys. Electron. Atom. Coll. (New York, 1989) p. 205

7.53 G.H. Wannier: Phys. Rev. **90**, 817 (1953)

7.54 U. Fano: Phys. Rev. A **22**, 2660 (1980)

7.55 J.M. Feagin: Proc. 15th Int. Conf. Phys. Electron Atom. Coll. (Brighton, 1987) p. 749

7.56 D.S. Newman, S.J. Buckman: Proc. 16th Int. Conf. Phys. Electron. Atom. Coll. (New York, 1989) p. 190

7.57 Y. Komninos, C.A. Nicolaides: J. Phys. B **19**, 1701 (1986)

7.58 D.W.O. Heddle: Proc. Roy. Soc. Lond. A **352**, 441 (1977)

7.59 F.H. Read: J. Phys. B **23**, 951 (1990)

7.60 D.W.O. Heddle, R.G.W. Keesing, J.M. Kurepa: Proc. Roy. Soc. A **334**, 135 (1973)

7.61 U. Fano, J.W. Cooper: Phys. Rev. A **137**, 1364 (1965)

7.62 P.J.M. van der Burgt, J. van Eck, H.G.M. Heideman: J. Phys. B **19**, 2015 (1986)

7.63 H.G.M. Heideman: "The Formation and Decay of Triply Excited He$^-$ States in e–He Scattering", in *Multiphoton Processes*, Proc. 4th Int. Conf. Multiphoton Processes, ed. by S.J. Smith, P.L. Knight (Cambridge Univ. Press 1988)

7.64 S. Cvejanović: Invited Papers of 15th Int. Conf. Phys. Electron. Atom. Coll. (Brighton, 1987) ed. by H.B. Gilbody, W.R. Newill, F.H. Read, A.C. Smith (Amsterdam 1988) p. 767
7.65 P. Mitchell, J. Baxter, J. Comer: J. Phys. B **13**, 4481 (1980)
7.66 D. Spence: J. Phys. B **14**, 129 (1981)
7.67 H.W. Dassen, R. Gomez, G.C. King: J. Phys. B **16**, 1481 (1983)
7.68 N.S. Scott, K.L. Bell, P.G. Burke, K.T. Taylor: J. Phys. B **15**, L627 (1982)
7.69 S.J. Buckmann, P. Hammond, G.C. King, F.H. Read: J. Phys. B **16**, 4219 (1983)
7.70 U. Fano, C.D. Lin: in *Atomic Physics 4*, ed. by G. zu Putlitz, E.W. Weber, A. Winnaker (Plenum, New York 1975) pp. 47–70
7.71 F.H. Read: *Atomic Physics 7*, ed. by D. Kleppner (Plenum, New York 1981)
7.72 F.H. Read, J.N.H. Brunt, G.C. King: J. Phys. B **9**, 2209 (1976)
7.73 A.D. Baas, P. Hammond, S.J. Buckmann, G.C. King, F.H. Read: Proc. 15th Int. Conf. Electron Atom. Coll. (Brighton, 1987) p. 220
7.74 D.F. Dance, M.F.A. Harrison, A.C.H.A. Smith: Proc. Roy. Soc. A **290**, 74 (1966)
7.75 N.R. Daly, R.E. Powell: Phys. Rev. Lett. **19**, 1165 (1967)
7.76 S. Ormonde, W. Whitaker, L. Lipsky: Phys. Rev. Lett. **19**, 1161 (1967)
7.77 B. Peart, K.T. Dolder: J. Phys. B **6**, 2415 (1973)
7.78 A.I. Dashchenko, A.I. Imre, I.P. Zapesochny: Pis'ma Zh. Exp. Teor. Fiz. **19**, 223 (1974)/English transl.: Sov. Phys. JETP Lett. **40**, 249 (1975)
7.79 Ya.N. Semenyuk: Ukr. Fiz. Zh. **29**, 1252 (1984)
7.80 P.G. Burke, D.D. McVicar: Proc. Phys. Soc. **86**, 989 (1965)
7.81 P.G. Burke, D.D. McVicar, K. Smith: Phys. Rev. Lett. **11**, 559 (1963)
7.82 P.G. Burke, D.D. McVicar, K. Smith: Proc. Phys. Soc. **84**, 749 (1964)
7.83 P.G. Burke, D.D. McVicar, K. Smith: Phys. Lett. **12**, 215 (1964)
7.84 P.G. Burke, A.J. Taylor: Proc. Phys. Soc. **88**, 549 (1966)
7.85 P.G. Burke, A.J. Taylor: J. Phys. B **2**, 44 (1969)
7.86 D.H. Oza: Phys. Rev. A **35**, 4430 (1987)
7.87 M.I. Gaysak, V.I. Lengyel, V.T. Navrotsky, E.P. Sabad: Ukr. Fiz. Zh. **27**, 1617 (1982)
7.88 L. Lipsky, R. Anania, M.J. Conneely: Atomic Data and Nuclear Data Tables, **20**, 127 (1977)
7.89 D.R. Herrick, O. Sinagoglu: Phys. Rev. A **11**, 97 (1975)
7.90 S. Wakid, T. Callaway: Phys. Lett. A **78**, 137 (1980)
7.91 Y.K. Ho: J. Phys. B **12** 387 (1979)
7.92 V.S. Senashenko, A. Wague: J. Phys. B **12**, L269 (1979)
7.93 A. Wague, P.B. Ivanov, V.S. Senashenko: Vestn. Mosk. Univ. ser. Fiz. Astron., **23**, 49 (1982)
7.94 R.P. Madden, K. Codling: Astrophys. J. **141**, 364 (1965)
7.95 P. Dhez, D.L. Ederer: J. Phys. B **6**, L59 (1973)
7.96 P.R. Woodruff, I.A.R. Samson: Phys. Rev. A **25**, 848 (1982)
7.97 U. Fano: Phys. Rev. A **124**, 1866 (1961)
7.98 K.A. Berrington, A.E. Kingston, S.A. Salvini: J. Phys. B **16**, 2399 (1983)
7.99 C.A. Weatherford: J. Phys. B **9**, L135 (1976)
7.100 R.J.W. Henry: Phys. Repts. **68**, 1 (1981)
7.101 D.H. Oza: J. Phys. B **18**, L231 (1985)
7.102 D.H. Oza: Phys. Rev. A **33**, 824 (1986)
7.103 E.M. Karule, R.K. Peterkop: "Collisions of Slow Electrons with Atoms of Alkali Elements", in *Effective Cross Sections for Electron-Atom Collisions*, No3 (Zinatne, Riga 1965) pp. 3–31
7.104 D.L. Moores, D.W. Norcross: J. Phys. B **5**, 1482 (1972)
7.105 A.L. Sinfailam, R.K. Nesbet: Phys. Rev. A **7**, 1987 (1973)
7.106 B.P. Kaulakis: Opt. Spektrosk. **48**, 1047 (1980)/English transl.: Opt. Spectrosc. USSR **48**, 315 (1980)
7.107 I.I. Fabrikant: Opt. Spektrosk. **53**, 223 (1982)/English transl.: Opt. Spectrosc. USSR **53**, 131 (1982)
7.108 P.G. Burke, A.J. Taylor: J. Phys. B **2**, 859 (1969)
7.109 A.R. Johnston, P.D. Burrow: J. Phys. B **15**, 1745 (1982)

7.110 B. Stefanov: J. Phys. B **11**, L249 (1978)

7.111 D.L. Moores: J. Phys. B **9**, 1329 (1976)

7.112 L.M. Biberman, A.Kh. Mnatsakanyan, I.T. Yakubov: Usp. Fiz. Nauk **102**, 431 (1970)/English transl.: Sov. Phys.

7.113 O. Dulieu, C. Le Sech: Europhys. Lett. **3**, 975 (1987)

7.114 A.C. Fung, I.I. Matese: Phys. Rev. A **5**, 22 (1972)

7.115 S.M. Kazakov, V.I. Lengyel, A.E. Masalovitch, O.V. Khristoforov, V.T. Navrotsky, E.P. Sabad: Ukr. Fiz. Zh. **30**, 502 (1985)

7.116 O.I. Zatsarinny, V.I. Lengyel, E.P. Sabad, I.I. Cherlenyak: Izv. Akad. Nauk SSR, Ser. Fiz. **50**, 1377 (1986)

7.117 C.D. Lin: J. Phys. B **16**, 723 (1983)

7.118 B. Jaduszliver, A. Tino, B. Bederson: Phys. Rev. A **24**, 1249 (1981)

7.119 D. Leep, A. Gallagher: Phys. Rev. A **10**, 1082 (1974)

7.120 D.W. Norcross: Phys. Rev. Lett. **32**, 192 (1974)

7.121 M.A. Crees, M.J. Seaton, P.M.H. Wilson: Comput. Phys. Comm. **15**, 23 (1978)

7.122 I.I. Cherlenyak, S.M. Kazakov, O.V. Khristoforov, V.I. Lengyel, E.A. Masalovitch, E.P. Sabad: 14th Int. Conf. Phys. Electron. Atom. Coll. (Palo Alto, 1985) p. 693

7.123 F.M. Pichanick, J.A. Simpson: Phys. Rev. A **5**, 64 (1968)

7.124 J.R. Gibson, K.T. Dolder: J. Phys. B **2**, 741 (1969)

7.125 I.I. Cherlenyak, V.I. Lengyel, E.P. Sabad: Acta Phys. Hungarica **63**, 373 (1988)

7.126 G.F. Drukarev: *Collisions of Electrons with Atoms and Molecules* (Nauka, Moscow 1978) (in Russian)

7.127 J. Perel, P. Englander, B. Bederson: Phys. Rev. A **128**, 1448 (1962)

7.128 K. Rubin, B. Bederson, M. Goldstein, R.E. Collins: Phys. Rev. A **182**, 201 (1960)

7.129 S.M. Kazakov, O.V. Khristoforov: Proc. 14th Int. Conf. Phys. Electron. Atom. Coll. (Palo Alto, 1985) p. 144

7.130 V.I. Lengyel, E.P. Sabad, I.I. Cherlenyak, S.M. Kazakov, O.V. Khristoforov: Opt. Spektrosk. **65**, 1032 (1988)/English transl.: Opt. Spectrosc. USSR

7.131 G.H. Green: Phys. Rev. A **23**, 661 (1981)

7.132 M. Eyb: J. Phys. B **9**, 101 (1976)

7.133 I. Simons: Int. J. Quant. Chem. **14**, 333 (1978)

7.134 P.G. Burke, J.F.B. Mitchell: J. Phys. B **6**, L161 (1973)

7.135 P.G. Burke, J.F.B. Mitchell: J. Phys. B **7**, 214 (1974)

7.136 M. Gehenn, E. Reichert: J. Phys. B **10**, 3105 (1977)

7.137 N.S. Scott, K. Bartshat, P.G. Burke, W.B. Eisner: J. Phys. B **17**, L191 (1984)

7.138 N.S. Scott, K. Bartschat, P.G. Burke, O. Nogy, W.B. Eisner: J. Phys. B **17**, 3775 (1984)

7.139 U. Fano: Repts. Progr. Phys. **46**, 97 (1983)

7.140 K. Scheibner, A. Hazi, R.J.W. Henry: Phys. Rev. A **35**, 48 (1987)

7.141 R.G. Athay: Astrophys. J. **146**, 223 (1966)

7.142 V.V. Zhukov, V.S. Kucherov, E.L. Latush, M.F. Sém: Kvant. Elektr. **4**, 1257 (1977)/English transl.: J. Quantum·Electron. USSR **7**, 708 (1977)

7.143 G.G. Petrash: Usp. Fiz. Nauk. **105**, 645 (1971)/English transl.: Sov. Phys.-Usp. **14**, 747 (1972)

7.144 L.P. Presnyakov, A.M. Urnov: Zh. Exp. Teor. Fiz. **68**, 61 (1975)/English transl.: Sov. Phys. JETP **41**, 31 (1975)

7.145 V.A. Bazylev, M.I. Chibisov: Usp. Fiz. Nauk **133**, 617 (1981)/English transl.: Sov. Phys.-Usp. **24**, 276 (1981)

7.146 I.P. Zapesochny, A.I. Dashchenko, V.I. Frontov, A.I. Imre, A.M. Gomonay, V.I. Lengyel, V.T. Navrotsky, E.P. Sabad: Pis'ma Zh. Exp. Teor. Fiz. **39**, 45 (1984)/English transl.: Sov. Phys. JETP Lett. **39**, 51 (1984)

7.147 P.O. Taylor, R.A. Phaneuf, G.H. Dunn: Phys. Rev. A **22**, 435 (1980)

7.148 V.A. Kel'man, A.I. Dashchenko, I.P. Zapesochny, A.I. Imre: Dokl. Akad. Nauk SSSR **220**, 65 (1975)/English transl.: Sov. Phys. - Dokl. **20**, 38 (1975)

7.149 P.O. Taylor, G.H. Dunn: Phys. Rev. A **8**, 2304 (1985)

7.150 V.I. Frontov: Opt. Spektrosk. **59**, 460 (1975)/English transl.: Opt. Spectrosc. USSR **59**, 277 (1985)

7.151 V.I. Frontov, A.I. Dashchenko, A.I. Imre, I.P. Zapesochny: Ukr. Fiz. Zh. **30**, 883 (1985)

7.152 V.I. Lengyel, V.T. Navrotsky, E.P. Sabad: J. Phys. B **23**, 1 (1990); Ukr. Fiz. Zh. **28**, 1798 (1983)

7.153 P.G. Burke, D.L. Moores: J. Phys. B **1**, 575 (1968)

7.154 I. Mitroy, D.G. Griffin, D.W. Norcross, M.S. Pindzola: Phys. Rev. A **38**, 3339 (1988)

7.155 C. Mendoza: J. Phys. B **14**, 2465 (1981)

7.156 V.I. Lengyel, E.A. Masalovitch, V.T. Navrotsky, E.P. Sabad: Izv. VUZov Fiz. **28**, 25 (1985)

7.157 M.A. Hayes, D.W. Norcross, J.B. Mann, D.W. Robb: J. Phys. B **10**, L429 (1977)

7.158 M.Blaha: Aston. Astrophys. **16**, 435 (1972)

7.159 O.I. Zatsarinny, V.I. Lengyel, E.A. Masalovich: Opt. Spektrosk. **67**, 20 (1989)/English transl.: Opt. Spectrosc. USSR

7.160 V.I. Lengyel, V.T. Navrotsky, E.P. Sabad: Izv. VUZov Fiz. **27**, 23 (1984)

7.161 D.L. Moores: Proc. Phys. Rev. **91**, 830 (1967)

7.162 D.W. Norcross, M.J. Seaton: J. Phys. B **9**, 2983 (1976)

7.163 P.L. Altick: Phys. Rev. A **169**, 21 (1968)

7.164 G. Mehlman-Ballofet, J.M. Esteva: Astrophys. J. **157**, 945 (1969)

7.165 J.M. Esteva, G. Mehlman-Ballofet: J. Quantum Spectrosc. Radiation Transfer **12**, 1291 (1977)

7.166 L. Johansson: Arkiv. Fiz. **23**, 119 (1963)

7.167 J. Mitroy, D.W. Norcross: Phys. Rev. A **37**, 3755 (1988)

7.168 F.A Parpia, D.W. Norcross, F.J. da Paixao: Phys. Rev. A **34**, 4777 (1986)

7.169 V.I. Lengyel, E.A. Masalovich, V.T. Navrotsky, E.P. Sabad: Proc. XVI Int. Conf. Phys. Electron. Atom. Coll. (New York, 1989) p. 98

7.170 C. Nicolaides: Phys. Rev. A **6**, 2078 (1972)

7.171 J. Callaway: Comments. Atom. Molec. Phys. **10**, 279 (1981)

7.172 O.I. Zatsarinny, V.I. Lengyel, V.T. Navrotsky, E.P. Sabad: Opt. Spektrosk. **56**, 585 (1984)/English transl.: Opt. Spectrosc. USSR

7.173 R. Bonanno, C. Clark, T. Lucatorto: Phys. Rev. A **34**, 2082 (1986)

7.174 C. Mendoza: J. Phys. B **14**, 397 (1981)

7.175 M.A. Baig, J.P. Connerade: Proc. Roy. Soc. A: **364**, 353 (1978)

7.176 Th. M. El-Sherbini, A.A. Rahman: Ann. Phys. (Leipzig) **39**, 333, (1982)

7.177 V. Pejčev, D. Rassi, T.W. Ottley: J. Phys. B **10**, 2913 (1977)

7.178 T.N. Chang: Phys. Rev. A **136**, 5468 (1987)

7.179 V.I. Lengyel, V.T. Navrotsky, E.P. Sabad: Opt. Spektrosk. **58**, 760 (1985)/English transl.: Opt. Spectrosc. USSR

7.180 V.I. Lengyel, V.T. Navrotsky, E.P. Sabad: Ukr. Fiz. Zh. **29**, 1484 (1984)

7.181 V. Pejčev, D. Rassi, K.J. Ross: J. Phys. B **13**, L305 (1980)

7.182 M.J. Seaton: J. Phys. B **7**, 1817 (1974)

7.183 O.I. Zatsarinny, V.I. Lengyel, E.A. Masalovich: Proc. XVI Int. Conf. Phys. Electron. Atom. Coll. (New York, 1989) p. 369. See also: Phys. Rev., **44**, 7287 (1991)

7.184 A.I. Zapesochny, A.I. Imre, I.S. Aleksakhin, I.P. Zapesochny, O.I. Zatsarinny: Zh. Eksp. Teor. Fiz. **43**, 463 (1986)/English transl.: Sov. Phys. JETP **63**, 1155 (1986)

7.185 I.P. Zapesochny, A.I. Imre, E.E. Kontrosh, A.I. Zapesochny, A.N. Gomonay: Pis'ma Zh. Eksp. Teor. Fiz. **43**, 463 (1986)/English transl.: Sov. Phys. JETP Lett. **43**, 596 (1986)

7.186 I.I. Fabrikant: "Collisions of Slow Electrons with the Atoms of Alcaline Earth Elements", in *Atomic Processes* (Zinatne, Riga 1975) p. 80 (in Russian)

7.187 F.D. Burrow, J.A. Micheida, J. Comer: J. Phys. B **9**, 3225 (1976)

7.188 N.I. Romanyuk, O.B. Shpenik, I.P. Zapesochny: Pis'ma Zh. Eksp. Teor. Fiz. **32**, 472 (1980)/English transl.: Sov. Phys. JETP Lett.

7.189 V.I. Kelemen, E.Yu. Remeta, E.P. Sabad: Zh. Techn. Fiz. **61**, 46 (1991)

7.190 O.B. Shpenik, I.P. Zapesochny, L. Szóvtér, E.I. Nepijpov, N.I. Romanyuk, I.S. Aleksakhin: Proc. IX Int. Conf. Phys. Electron. Atom. Coll. (Paris, 1977) p. 1302. See also: Zh. Eksp. Teor. Fiz. **76**, 346 (1979)/English transl.: Sov. Phys. JETP **49**, 426 (1979)

7.191 J. Hunt, B.L. Moiseiwitsch: J. Phys. B **3**, 891 (1979)

7.192 H.A. Kurtz, Y. Ohrn: Phys. Rev. A **19**, 43 (1979)

7.193 J.K. van Blerkom: J. Phys. B **3**, 932 (1970)

7.194 M.Ya. Amusia, V.A. Sosnivker, N.A. Cherepkov, L.V. Chernysheva: *A Calculation of the Cross Section for Elastic Scattering of Slow Electrons by Ca Atoms*, Physicotechnical Institute of the USSR Academy of Sciences Rpt. N863, (Leningrad, 1983)

7.195 C. Froese-Fischer, J.B. Lagowsky, S.H. Vosko: Phys. Rev. Lett. **59**, 2263 (1987)

7.196 D.J. Pegg, J.S. Thompson, R.N. Compton, G.D. Alton: Phys. Rev. Lett. **59**, 2267 (1987)

7.197 G.F. Gribakin, B.V. Gul'tsev, V.K. Ivanov, M.Yu. Kuchiev: Proc. XVI Int. Conf. Phys. Electron. Atom. Coll. (New York, 1989) p. 851

7.198 D. Burrow, J. Comer: J. Phys. B **8**, L92 (1975)

7.199 I.P. Zapesochny, O.B. Shpenik: Zh. Eksp. Teor. Fiz. **50**, 890 (1966)/English transl.: Sov. Phys. JETP

7.200 O.B. Shpenik, I.P. Zapesochny, L. Szóvtér, E.E. Kontrosh, A.N. Zavilopulo: Zh. Eksp. Teor. Fiz. **65**, 1979 (1973)/English transl.: Sov. Phys. JETP

7.201 I.P. Zapesochny, L. Szóvtér, O.B. Shpenik, A.N. Zavilopulo: Dokl. Akad. Nauk SSSR **214**, 1287 (1974)/English transl.: Sov. Phys. - Dokl.

7.202 J. Machugo, B. Palásthy, L. Szóvtér, G. Vitéz: Proc. XVI Int. Conf. Phys. Electron. Atom. Coll. (New York, 1989) p. 269

7.203 P.J. Hicks, S. Cvejanovič, J. Comer, F.H. Read, J.M. Sharp: Vacuum **24**, 573 (1974)

7.204 I.P. Zapesochny, O.B. Shpenik: dokl. Akad. Nauk SSSR **160**, 1053 (1965)/English transl.: Sov. Phys. - Dokl. **10**, 140 (1965)

7.205 C.E. Kuyatt, J.A. Simpson, S.R. Mielczarek: Phys. Rev. A **138**, 385 (1965)

7.206 U. Fano, J.W. Cooper: Phys. Rev. A **138**, 400 (1965)

7.207 T.W. Ottley, H. Kleinpoppen: J. Phys. B **8**, 621 (1975)

7.208 L. Koch, T. Heindor, E. Keichert: Z. Phys. A **316**, 127 (1984)

7.209 D.S. Newman, G.C. King, M. Zubek: Proc. XIII Int. Conf. Phys. Electron. Atom. Coll. (Berlin, 1983) p. 104

7.210 K. Albert, C. Christian, T. Heindorff, E. Reichert, S. Schön: J. Phys. B **10**, 3733 (1977)

7.211 D.W.O. Heddle: J. Phys. B **11**, L711 (1978)

7.212 S.M. Kazakov, A.I. Korotkov, O.B. Shpenik: Zh. Eksp. Teor. Fiz. **78**, 1687 (1980)/English transl.: Sov. Phys. JETP **51**, 847 (1980)

7.213 N.S. Scott, P.G. Burke, K. Bartshat: J. Phys. B **16**, L361 (1983)

7.214 H. Kleinpoppen: Phys. Rev. A **3**, 2015 (1971)

7.215 G.F. Hanne: *Coherence and Correlation in Atomic Physics*, ed. by H. Kleinpoppen, J. Williams (Plenum, New York 1980)

7.216 S.M. Khalid, H. Kleinpoppen: J. Phys. B **17**, 243 (1984)

7.217 T.T. Gien: J. Phys. B **15**, 4617 (1982)

7.218 J. Kessler: *Polarized Electrons* (Springer, Berlin, New York 1976, second ed. Heidelberg 1991)

7.219 G.F. Drukarev, V.D. Obyedkov: Usp. Fiz. Nauk **127**, 621 (1979)/English transl.: Sov. Phys. - Usp. **22**, 236 (1979)

7.220 N.B. Delone, M.V. Fedorov: Usp. Fiz. Nauk **127**, 651 (1979)/English transl.: Sov. Phys. - Usp. **22**, 252 (1979)

7.221 D.T. Pierce, R.J. Celotta, G.C. Wang: Rev. Sci. Instrum. **51**, 478 (1980)

7.222 G.F. Hanne, J. Kessler: J. Phys. B **9**, 791 (1976)

7.223 G. Baum, E. Kisker, E. Raith, W. Schröder, U. Sillmen, D. Zenses: J. Phys. B **14**, 4377 (1981)

7.224 G. Baum, L. Frost, W. Raith, U. Sillmen: J. Phys. B **32**, 1667 (1989)

7.225 G. Baum, M. Moede, W. Raith, U. Sillmen: Proc. XIV Int. Conf. Phys. Electron. Atom. Coll. (Palo Alto, 1985) p. 182

7.226 V.I. Lengyel, V.T. Navrotsky, E.P. Sabad: Usp. Fiz. Nauk **151**, 425 (1987)/English transl.: Sov. Phys. - Usp. **30**, 220 (1987)
7.227 D.L. Moores: J. Phys. B **19**, 1843 (1986)
7.228 W. Jitschin, S. Osimitsch, H. Reihe, H. Kleinpoppen: Proc. XIII Int. Conf. Phys. Electron. Atom. Coll. (Berlin, 1983) p. 150
7.229 A. Wolcke, K. Bartshat, H. Borgmann, G.F. Hanne, J. Kessler: J. Phys. B **16**, 639 (1983)
7.230 I.I. Shafranyosh, T.A. Snegurskaja, I.S. Aleksakhin: Proc. XVI Int. Conf. Phys. Electron. Atom. Coll. (New York, 1989) p. 196
7.231 I.S. Aleksakhin, I.I. Shafranyosh: Opt. Spektrosk. **42**, 773 (1977)/English transl.: Opt. Spectrosc. USSR
7.232 M. Chen, B. Crasemann: Phys. Rev. A **37**, 2886 (1989)
7.233 K.J. Reed, M.H. Chen, A. Hazi: Phys. Rev. A **36**, 3117 (1987)
7.234 K.J. Reed, M.H. Chen: Proc. XVI Int. Conf. Phys. Electron. Atom. Coll. (New York, 1989) p. 375

Chapter 8

8.1 B.M. Smirnov: *Excited Atoms* (Atomizdat, Moscow 1982) (in Russian)
8.2 E.V. Aglitsky, U.I. Safronova: *Spectroscopy of Autoionization States of Atomic Systems* (Energoatomizdat, Moscow 1988) (in Russian)
8.3 N.B. Delone, V.P. Krainov: *An Atom in an Intense Optical Field* (Energoatomizdat, Moscow 1984) (in Russian) / English transl.: Atoms in Strong Light Fields (Springer, Heidelberg 1985)
8.4 P. Lambropulos: Multiphoton Processes (Springer, Heidelberg 1991)
8.5 L.A. Wainshtein, I.I. Sobel'man, E.A. Yukov: *Excitation of Atoms and Spectral Line Broadening* (Nauka, Moscow 1979) (in Russian) / English transl.: Excitation of Atoms and Broadening of Spectral Lines (Springer, Heidelberg 1985)
8.6 I.I. Sobelman: Atomic Spectra and Radiative Transitions (Springer, Heidelberg 1991) ·
8.7 R.K. Janev, L.P. Presnyakov, V.P. Shevelko: Physics of Highly Charged Ions. (Springer, Heidelberg 1985)
8.8 V.M Galitsky, E.E. Nikitin, B.M. Smirnov: *Theory of Collisions of Atomic Particles* (Nauca, Moscow 1981) (in Russian)
8.9 V.I. Lengyel, M.I. Haysak: Adv. At. Mol. Phys. **27**, 245 (1990)
8.10 E.A. Perelstein, G.D. Shirkov: *Electron-Ion Ringe Phenomena and the Problem of Highly Charged Ion Production*, JINR Report No. E9-85-4, Dubna (1985)
8.11 W. Sandhas: Proc. XII Europ. Few-Body Conf. (Uzhgorod, 1990) p. 350

Subject Index